VOLUME EIGHT

ANNUAL REPORTS IN
COMPUTATIONAL
CHEMISTRY

VOLUME EIGHT

ANNUAL REPORTS IN
COMPUTATIONAL CHEMISTRY

Edited by

Ralph A. Wheeler

Department of Chemistry and Biochemistry,
Duquesne University,
600 Forbes Avenue,
Pittsburgh,
PA 15282-1530.

Sponsored by the Division of Computers in Chemistry
of the American Chemical Society

ELSEVIER Amsterdam • Boston • Heidelberg • London • New York • Oxford
Paris • San Diego • San Francisco • Singapore • Sydney • Tokyo

Elsevier

Radarweg 29, PO Box 211, 1000 AE Amsterdam, The Netherlands
The Boulevard Langford Lane, Kidlingfon, Oxford, OX51GB, UK

First edition 2012

Library of Congress Cataloging-in-Publication Data
A catalogue record for this book is available from the Library of congress

British Library Cataloging in Publication Data
A catalogue record for this book is available from the British Library

ISBN: 978-0-444-59440-2
ISSN: 1574-1400

For information on all Elsevier publications visit
our website at store.elsevier.com

Printed and bound in USA
12 13 14 15   10 9 8 7 6 5 4 3 2 1

Working together to grow
libraries in developing countries

www.elsevier.com | www.bookaid.org | www.sabre.org

ELSEVIER   BOOK AID   Sabre Foundation
              International

CONTENTS

Section B: Biological Modeling

Nathan Baker

Section C: Bioinformatics

Wei Wang

CONTRIBUTORS

Qin Cai
Department of Biomedical Engineering, University of California, Irvine, CA, USA;
Department of Molecular Biology and Biochemistry, University of California, Irvine,
CA, USA

David A. Dixon
Department of Chemistry, The University of Alabama, Tuscaloosa, Alabama, USA

David Feller
Department of Chemistry, Washington State University, Pullman, Washington, USA

Mark R. Hoffmann
Chemistry Department, University of North Dakota, Grand Forks, North Dakota, USA

Meng-Juei Hsieh
Department of Molecular Biology and Biochemistry, University of California, Irvine,
CA, USA

Wanyi Jiang
Department of Chemistry, Center for Advanced Scientific Computing and Modeling
(CASCaM), University of North Texas, Denton, Texas, USA

Dian Jiao
Center for Biological and Material Sciences, Sandia National Laboratories, Albuquerque,
New Mexico, USA

Yuriy G. Khait
Chemistry Department, University of North Dakota, Grand Forks, North Dakota, USA

Guohui Li
Molecular Modeling and Design, State Key Laboratory of Molecular Reaction
Dynamics, Dalian Institute of Chemical Physics, Chinese Academy of Sciences, Dalian,
P.R. China

Ray Luo
Department of Molecular Biology and Biochemistry, University of California, Irvine,
CA, USA

Kirk A. Peterson
Department of Chemistry, Washington State University, Pullman, Washington, USA

Lawrence R. Pratt
Department of Chemical & Biomolecular Engineering, Tulane University, New Orleans,
Louisiana, USA

Susan B. Rempe
Center for Biological and Material Sciences, Sandia National Laboratories, Albuquerque,
New Mexico, USA

Pengyu Ren
Department of Biomedical Engineering, The University of Texas at Austin, Texas, USA

David M. Rogers
Center for Biological and Material Sciences, Sandia National Laboratories, Albuquerque, New Mexico, USA

Hujun Shen
Molecular Modeling and Design, State Key Laboratory of Molecular Reaction Dynamics, Dalian Institute of Chemical Physics, Chinese Academy of Sciences, Dalian, P.R. China

Jun Wang
Department of Molecular Biology and Biochemistry, University of California, Irvine, CA, USA

Angela K. Wilson
Department of Chemistry, Center for Advanced Scientific Computing and Modeling (CASCaM), University of North Texas, Denton, Texas, USA

Zhen Xia
Department of Biomedical Engineering, The University of Texas at Austin, Texas, USA

Xiang Ye
Department of Molecular Biology and Biochemistry, University of California, Irvine, CA, USA

PREFACE

Annual Reports in Computational Chemistry (ARCC) focuses on providing concise, timely reviews of topics important to researchers in computational chemistry. *ARCC* is published and distributed by Elsevier and sponsored by the American Chemical Society's Division of Computers in Chemistry (COMP). All members in good standing of the COMP Division receive a copy of the *ARCC* as part of their member benefits. The Executive Committee of the COMP Division is very pleased that previous volumes have received an enthusiastic response from our readers and we expect to continue the tradition by delivering similarly high-quality volumes of *ARCC*. To ensure that you receive future installments of this series, please join the Division as described on the COMP website at http://www.acscomp.org.

Volume 8 is smaller than previous volumes and is tightly focused on new computational methods. Section A (Quantum Mechanics, edited by Gregory Tschumper) leads off with a review of first principles thermochemistry and nicely introduces the following paper discussing potential energy surfaces and electronic excited states. The next contribution describes subsystem Kohn–Sham density functional theory and thus introduces some considerations necessary for the first paper of Section B (Biological Modeling, edited by Nathan Baker), model thermodynamics of hydration. This paper, in turn, leads into the next two papers, a review of physics-based, coarse-grained potentials for proteins, and, in Section C (Bioinformatics, edited by Wei Wang) a contribution describing Poisson–Boltzmann implicit solvation models. To provide easy identification of past reports, we plan to continue the practice of cumulative indexing of both the current and past editions.

The current and past volumes of *Annual Reports in Computational Chemistry* have been assembled entirely by volunteers to produce a high-quality scientific publication at the lowest cost possible. The Editor and the COMP Executive Committee extend our gratitude to the many people who have given their time so generously to make this edition of *Annual Reports in Computational Chemistry* possible. The authors of each of this year's contributions and the Section Editors have graciously dedicated significant amounts of their time to make this volume successful. This year's edition could not have been assembled without the help of Shellie Bryant and Paul

Milner of Elsevier. Thank you one and all for your hard work, your time, and your contributions.

We trust that you will continue to find ARCC to be interesting and valuable. We are actively planning the ninth volume and welcome input from our readers about future topics, so please contact the Editor to make suggestions and/or to volunteer as a contributor.

Sincerely,
Ralph A. Wheeler, Editor

Quantum Chemistry

Section Editor: Gregory S. Tschumper

Department of Chemistry and Biochemistry, University of Mississippi, University, MS, USA

CHAPTER ONE

A Practical Guide to Reliable First Principles Computational Thermochemistry Predictions Across the Periodic Table

David A. Dixon*,1 David Feller†, Kirk A. Peterson†

*Department of Chemistry, The University of Alabama, Tuscaloosa, Alabama, USA
†Department of Chemistry, Washington State University, Pullman, Washington, USA
1Corresponding author: E-mail: dadixon@bama.ua.edu

Contents

Abstract

The Feller–Peterson–Dixon approach to the reliable prediction of thermochemical properties to chemical accuracy is described. The method calculates the total atomization energy of a molecule and then uses experimental atomic heats of formation to calculate the heat of formation. The method is based on extrapolating coupled cluster CCSD(T) calculations of the valence electronic energy to the complete basis set limit followed by additive corrections to the electronic energy including: core–valence interactions, scalar relativistic, spin–orbit, and higher order corrections beyond CCSD(T). Corrections for the nuclear motion in terms of the zero-point energy need to be included and for high-accuracy, Born–Oppenheimer corrections may be added. Issues with these terms and the experimental atomic data are described. A summary of the reliability of the approach is presented.

Annual Reports in Computational Chemistry, Volume 8
ISSN 1574-1400,
http://dx.doi.org/10.1016/B978-0-444-59440-2.00001-6

1

1. INTRODUCTION

Computational chemistry can now be used to provide quantitatively accurate thermochemical properties for a wide range of systems. Dramatic changes in both accessible computing power and developments in approaches to solving the electronic Schrödinger equation have occurred at the same time. In addition, these methods have been implemented in easy-to-use and widely available software packages that perform well on modern computer architectures. The focus here is on correlated molecular orbital theory methods, e.g. wave function-based approaches as they represent a scheme that can systematically approach the exact solution of the molecular electronic Schrödinger equation (full configuration interaction) in a given basis set. This is known as the *n*-particle approximation. We have chosen to use coupled cluster theory for our correlated molecular orbital treatment [1]. The other piece needed for high accuracy is the choice of the basis set used to describe the underlying molecular orbitals. This is known as the 1-particle approximation and it strongly influences the accuracy of a quantum chemistry calculation. The introduction of the systematically convergent [to the complete basis set (CBS) limit] correlation-consistent basis sets [2] in 1989, cc-pV*n*Z where *n* = D, T, Q, 5, etc., has led to a paradigm shift in strategies for accurate solutions of the electronic Schrödinger equation. Using sequences of these basis sets with a chosen method, one can estimate the CBS limit, thus eliminating the 1-particle truncation problem arising from basis set incompleteness. Not only does this provide us with a measure of the intrinsic error associated with a given method but also it allows us to decouple the 1-particle and n-particle expansions in a way that was previously impossible. In addition, issues with basis set superposition error are minimized. Peterson and co-workers have extended these basis sets beyond the main group elements to the transition metals. These basis sets have recently been reviewed by Peterson in this series [3].

Our goal is to design a practical computational chemistry approach that provides "chemical accuracy" in terms of thermochemistry. Chemical accuracy is usually interpreted in the thermochemistry literature as ± 1 kcal/mol (± 4.184 kJ/mol). Attempting to achieve reliable, consistent chemical accuracy for a general polyatomic molecule from a single calculation has proven to be an impossible task at this point in the evolution of quantum chemistry techniques. One is confronted by the need to recover significant amounts of electron correlation while simultaneously using a 1-particle basis set near the CBS limit for the energy of all the electrons. In addition, the geometry and vibrational frequencies must also be treated at a high level to achieve chemical accuracy. The issue in performing such calculations is the computational scaling of the method with the size of the basis set. The lowest level n-particle correlation method that can be used to obtain near chemical accuracy for thermochemical properties is the CCSD(T) (coupled cluster singles and doubles with perturbative triples) method [4,5,6], which computationally scales as $n^3 N^4$, where n is the number of occupied orbitals and N the number of virtual orbitals. This is usually given as M^7, where M is the number of basis functions, although for a fixed molecule the cost actually scales as just N^4. To directly achieve chemical accuracy with CCSD(T) without extrapolation would require correlation of both the valence and outer-core electrons with a basis set of at least aug-cc-pCV6Z or aug-cc-pCV7Z quality [7]. This is currently possible for only very small molecules, especially as one has to calculate both the equilibrium geometry and the vibrational frequencies.

The general solution to this computational problem is to use a composite approach, which is based on the concept that smaller components are additive and that these terms can be most efficiently calculated at different levels of theory (but comparable levels of accuracy) to reduce the computational cost. We are interested in predicting the heat of formation as a key thermodynamic quantity. The issue with calculating heats of formation directly from the total energy of a molecule and its atoms is that the standard state for the heat of formation corresponds to the elements in their most stable form at 298 K. This choice raises issues with the calculations of total energies since these standard states often do not correspond to atomic gas-phase species. Thus, our approach is to calculate the total atomization energy and account for all of the terms needed to obtain this quantity as accurately as possible. As a consequence, securing high accuracy in the heat of formation depends upon achieving accuracy in the total atomization energy and being able to combine that with accurate experimental heats of formation of the atoms at 0 K. The expression for the zero-point inclusive, total atomization energy ($TAE = \Sigma D_0$) is given by the following:

$$TAE = \sum D_0$$
$$= \Delta E_{elec}(CBS) + \Delta E_{CV} + \Delta E_{SR} + \Delta E_{SO} + \Delta E_{HO} + \Delta E_{ZPE}$$
$$(1.1)$$

The various energy terms in the calculation are treated as accurately as possible, exploiting the fact that each piece can be handled with different levels of approximation. The first term in the energy expression is the electronic energy contribution of the "valence electrons," which we will take to be the CBS limit of the CCSD(T) energy. For molecular systems with wave functions dominated by a single Slater determinant, this is an appropriate approach. The second term is the correlation of the outer-core electrons with themselves and with the valence electrons, the core–valence (CV) term. If possible, this contribution is also estimated at the CBS limit. The third and fourth terms deal with the effects of relativity with the third term accounting for scalar relativistic effects and the fourth term dealing with the spin–orbit (SO) interaction. The fifth term deals with corrections to the TAE beyond the CCSD(T) level. The final value is the zero-point energy, which will be negative and decrease the TAE. The TAE, i.e. the enthalpy for reaction (2), can then be used to calculate the heat of formation of the molecule $A_aB_bC_c$ at 0 K from reaction (1.2) with expression (1.3).

$$A_aB_bC_c \rightarrow aA + bB + cC \qquad (1.2)$$

$$\Delta H_f^{0K}(A_aB_bC_c) = a\Delta H_f^{0K}(A) + b\Delta H_f^{0K}(B) + c\Delta H_f^{0K}(C) - \sum D_0 \qquad (1.3)$$

2. COMPUTATIONAL APPROACH TO RELIABLE THERMOCHEMICAL PREDICTIONS

We will describe how to deal with each of the terms in Eqn (1.1) in more detail below after a discussion of the Feller–Peterson–Dixon (FPD) approach [8,9]. Our composite approach is built on an extensive set of studies over about 15 years from our research group [10,11,12,13,14]. In addition to these leading references, many more examples are noted in the other references in this chapter. Before describing our approach, we note that there are a variety of other prescriptions using similar procedures. In 1989, the year that the correlation-consistent basis sets were first reported, Pople and co-workers reported the Gaussian-1 (G1) composite procedure for the calculation of total atomization energies, ionization potentials, as well as proton and electron affinities [15,16]. This was followed by the G2 [17], G3 [18], and G4 [19,20] composite theories, which were in turn followed by a large number of variants based on different reduced correlation levels and different treatments of the geometry and zero-point energies. These methods have been summarized in a recent review [8]. A key feature of the Gn methods is that beyond G2, they are parameterized to a set of experimental data (e.g. the G3/05 test set of 454 experimental energy differences) to deal with deficiencies in many of the approximations to Eqn (1.2) leading to the inclusion of an empirically based higher level correction. Petersson and co-workers have developed a number of procedures for estimating the correlation energy at the CBS limit resulting in the CBS models [21,22,23,24,25]. Both the Gn and Petersson CBS methods can be readily accessed in the Gaussian programs by a single keyword as can the W1 method noted next. Martin and co-workers introduced the Weizmann-n "black box" composite methods and have extended them from W1 to W4, also with many variants [26,27,28,29]. A number of groups have contributed to the development of the High-accuracy Extrapolated Ab Initio Thermochemistry (HEAT) model, which was intended to "achieve high accuracy for enthalpies of formation of atoms and small molecules" without resorting to empirical scale factors [30]. Alternate implementations of HEAT have also been reported [31,32]. Using an approximation to the FPD approach to be described in more detail, Wilson and co-workers developed an MP2-based procedure that included a CCSD(T) or multireference component. Their correlation-consistent composite approach (ccCA) has been applied to a range of elements including first-row transition metals [33,34,35,36,37,38]. A multireference analogue of ccCA, denoted as MR-ccCA, is described in Chapter 2 of this volume.

2.1. Calculating ΔE_{elec}(CBS)

The ΔE_{elec} term in Eqn (1.1) represents the largest single contribution to the atomization energy in all high-level approaches to computational thermochemistry. We normally choose to use the single reference-based CCSD(T) method as our "gold standard" for the treatment of the valence [frozen core (FC)] correlation energy. It has proven to be successful even in cases where the molecule's wave function has significant multireference character. However, in situations where a molecule's wave function is strongly multiconfigurational, we would replace CCSD(T) with complete active space self-consistent field configuration interaction (CAS-CI) theory [39]. For open-shell calculations, there are different approaches to the solution of the CCSD(T) equations and we recommend the use of the restricted method for the starting Hartree–Fock (HF) wave function and then relaxing the spin restriction in the coupled cluster portion of the calculation. This method is conventionally labeled R/UCCSD(T) [40]. An estimate of the potential for significant multireference character in the wave function can be obtained from the T_1 [41] and D_1 [42,43] diagnostic for the CCSD calculation.

If the highest level of accuracy is being sought in $\Delta E_{elec}(CBS)$, it is necessary to factor in the effect of geometry optimization at the CCSD(T) or higher level of theory. In the FPD approach, we would perform FC CCSD(T) geometry optimizations with a sequence of the correlation-consistent basis sets in an effort to evaluate energy differences at or near the CBS limit geometries. However, reasonable results can be obtained with lower level geometries and one should try to provide the best geometry possible for each level of basis set.

By exploiting the systematic convergence properties of the correlation-consistent basis sets including additional diffuse functions [2,44,45], it is possible to obtain accurate estimates of the CBS limit without having to resort to such extremely large basis sets that would unavoidably limit our composite approach to small diatomic molecules. These basis sets were initially only available for first-, second-, and third-row main group elements. The high atomic numbers (Z) of the heavier elements beyond the second row imply that relativistic effects must be properly included to attain even semiquantitative accuracy [46,47]. For many of the properties of interest, the core electrons are chemically inert. It is possible to eliminate the core contributions from direct consideration by using small-core pseudo-potentials or relativistic effective core potentials (RECPs) [48,49]. This is equivalent to ignoring the correlation effects of the 1s electrons for second-row atoms like S and Cl. Recently, Peterson and co-workers [50] have developed correlation-consistent basis sets in combination with effective core potentials from the Stuttgart/Köln group for all of the main group atoms, thereby enabling reliable predictions for compounds containing heavier elements. A small-core RECP should be used for the third-row and higher main group elements. The RECP for the fourth row subsumes the $(1s^2, 2s^2, 2p^6, 3s^2, 3p^6, 3d^{10})$ orbital space into the 28-electron core set, leaving the $(4s^2, 4p^6, 5s^2, 4d^{10}$ and $5p^x)$ space to be handled explicitly. In a normal valence correlation treatment, only the $(5s^2, 5p^x)$ electrons would be active. For the third row, a similar basis set can be used with the $(1s^2, 2s^2, 2p^6)$ electrons in the core and the remaining electrons are handled explicitly. The basis sets are designed such that only the spherical component subset (e.g. 5-term d functions, 7-term f functions, etc.) of the Cartesian polarization functions should be used. Peterson and co-workers [51] have recently developed such aug-cc-pVnZ-PP basis sets for the transition metals as well, and we recommend them in their small-core form. Good sources for the basis sets are the EMSL Basis Set Exchange web site [52,53,54] and the K. Peterson web site at Washington State University [55]. Many computer programs have basis sets embedded in them but one should take care to make sure that the basis sets are implemented the same in different programs and even in different versions of the same program when calculating various quantities.

The uniform, systematic convergence properties of the correlation-consistent basis sets coupled with the steep rise in computational costs with increasing basis set size have motivated a variety of simple extrapolation formulas for estimating the CBS limit. Of these, we have tended to focus on a subset of five formulas [Eqns (1.4) [56,57], (1.5) [58], (1.6) [59,60], (1.7) [59], and (1.8) [61]], all of which have been recently reviewed [7]. These include:

$$E(n) = E_{CBS} + A^* \exp(-Bn) \tag{1.4}$$

$$E(n) = E_{CBS} + A \exp\left[-(n-1)\right] + B \exp\left[-(n-1)^2\right] \tag{1.5}$$

$$E(\ell_{max}) = E_{CBS} + B/\ell_{max}^3 \tag{1.6}$$

$$E(\ell_{max}) = E_{CBS} + A/(\ell_{max} + 1/2)^4 \tag{1.7}$$

$$E_{CBS} = (E_{n+1} - E_n)F_{n+1}^C + E_n, \tag{1.8}$$

where n is the basis set index, $n = 2$ (VDZ), 3 (VTZ), etc., E_{CBS} is the complete basis set energy, $E(n)$ is the energy with the nth basis set and A and B are fitting parameters. In Eqn (1.8), the parameters (F_{n+1}^C) are specific for the HF, CCSD and (T) pieces of the CCSD(T) energy. They were obtained by a least squares fitting procedure of seven reference molecules treated with large, uncontracted (s,p,d,f) basis sets. With the correlation-consistent basis set family, the basis set index, n, in Eqns (1.4) and (1.5), is identical to ℓ_{max} for second and third period elements Li–Ar. Some ambiguity may arise if the basis set index does not correspond to a single value of ℓ_{max} for all elements in the chemical system. Examples include simple systems involving hydrogen and transition metals with main group elements, such as C_2H_4 and ZnF. In such cases, we have chosen to use the highest ℓ_{max} value in the extrapolation formula. For a pure transition metal species, no ambiguity exists. In practice, our calculations on group IVB and VIB transition metal oxide clusters have shown that the effect of the choice of the value of n used with Eqn (1.5) is fairly small and one can use either the values from the main group compounds or from the transition metal as long as that choice is used consistently [62–64]. Eqns (1.6) and (1.7) are formally intended for just the correlation component of the total energy, with the HF component extrapolated separately or taken from the largest basis set value. In practice, the effect on energy differences of treating the HF component separately or extrapolating the total HF + correlation energy is small enough to be ignored with Eqns (1.4), (1.5) and (1.7). Equation (1.8) inherently separates these two contributions to the energy. Even with Eqn (1.6), which showed the largest impact due to extrapolating the total energy rather than the separate pieces, the effect was small because of the dominance of the change in correlation energy over that of the HF energy. Experience has shown that the "best" extrapolation formula varies with the level of basis set and the molecular system. Consequently, there is no universally agreed upon definition of the best extrapolation formula. For benchmarking purposes for smaller systems, one should go to the largest possible basis set. In practice, for most large molecules, the largest basis set that one can afford is aug-cc-pVQZ. A recent statistical comparison study of the effectiveness of CBS extrapolation formulas found that Eqn (1.5) produced the best results for the DTQ combination by a small margin [7]. This study involved 141 small-to-medium sized chemical systems (later expanded to 163 systems) composed of elements through the fifth period.

A more efficient alternative to the combination of conventional CCSD(T) and basis set extrapolation involves a fundamental technique to improve the basis set convergence rate at the correlated level, namely the use of explicitly correlated methods. The slow convergence of the correlation energy with basis set size is due to the poor description of the electron coalescence cusp by products of one-electron functions. Recently developed F12 methods[65–67] utilize an exponential correlation factor, $F_{12} = e^{-\gamma r_{12}}$, which, when combined with cc-pVnZ-F12 (n = D, T, Q) orbital and auxiliary basis sets [68,69], can yield results near aug-cc-pV(n + 2)Z quality with nearly a negligible increase in computer resources when the CCSD(T)-F12b method[70] is used. Our previous results for a collection of small hydrocarbons also indicated that extrapolation of F12 correlation energies[71] can be very accurate even with just a DZ/TZ pair of basis sets [72].

Care must be used in defining the valence electron space for many compounds. For example, it is well established that one has to include the $(n - 1)$s and $(n - 1)$p electrons in the alkali and alkaline earth elements especially in bonding to O and F, as the O and F valence orbitals can have energies below these. Thus, one should always check the eigenvalue spectrum of the occupied orbitals at the HF level to make sure that the proper electrons are being correlated. In addition, programmers have implemented different default valence space definitions in different programs. This issue can also arise in the main group compounds with an "inner" shell of d electrons, especially to the left side of the p-block for atoms such as Ga or In. In these cases, it may be necessary to include the d electrons in the valence space. We have also found that one may have to include the $(n - 1)$s and $(n - 1)$p electrons or the f electrons in the active space of the early transition metals. It has been shown that tight d functions are important for calculating accurate atomization energies for second-row elements [45]. One should include additional tight d functions in calculations for the second-row main group elements. The correlation-consistent basis sets containing extra tight d functions are denoted as aug-cc-pV$(n + d)Z$ in analogy to the original augmented basis sets. The other d orbital exponents were re-optimized for these basis sets.

As discussed in Section 1.2.2, in order to achieve consistently high accuracy, it is necessary to include the effect of the core electron correlation at some level. In a recent study of the heats of formation of the iodine and xenon fluorides, we found that it is very important to include the core electrons in the treatment of the correlation energy together with the appropriate weighted CV basis sets and then extrapolate these quantities to the CBS limit if accurate energies were to be determined [73,74]. This results in a substantially more expensive calculation due to the need to correlate many more electrons and include extra basis functions suitable for them. In this example, the core electrons with the aug-cc-pwCVnZ basis sets for at least D, T, Q were used in the extrapolation to the CBS limit. In addition, we have found that one has to go beyond the aug-cc-pVQZ basis set level for compounds such as H_2SO_4. Many of these types of issues arise when there is a large change in the formal oxidation state of an atom in the molecule to the dissociated atom. For example, in IF_7, there is a change in the formal oxidation state from +7 on the I in the molecule to 0 in the dissociated atom, and similarly in H_2SO_4, there is a change in the oxidation state from +6 on S in the molecule to 0 in the dissociated atom. An example of this effect is shown in Table 1.1. The results for the diatomic IF show that there is not much effect from the inclusion of CV correlation in the extrapolation. However, for the larger molecules, especially for IF_5 and IF_7, there is a clear effect. The experimental heats of formation (ΔH_f) of IF_5 and IF_7 are well established as -198.8 ± 0.4 kcal/mol at 0 K and -200.8 ± 0.4

Table 1.1 CCSD(T) atomization energies for IF$_x$ in kilocalories per mole [73]

Molecule	ΔE_{SO} molecule	ΔE_{SO} atom	$\sum D_0$ (0 K) (DTQ)	$\sum D_0$ (0 K) (Q5)	$\sum D_0$(0 K) (DTQ)$_{CV}$
IF $\rightarrow$ I + F	1.7	−7.63	63.28	63.05	63.30
IF$_2^+$ + e$^-$ $\rightarrow$ I + 2F	2.8	−8.02	−114.57	−114.96	−115.53
IF$_2^-$ $\rightarrow$ I + 2F + e$^-$	1.0	−8.02	205.62	205.34	205.13
IF$_3$ $\rightarrow$ I + 3F	2.2	−8.41	184.21	183.64	182.84
IF$_4^+$ + e$^-$ $\rightarrow$ I + 4F	2.1	−8.80	8.22	7.20	6.16
IF$_4^-$ $\rightarrow$ I + 4F + e$^-$	2.0	−8.80	338.12	337.30	337.00
IF$_5$ $\rightarrow$ I + 5F	2.3	−9.19	320.97	319.70	318.61
IF$_6^+$ + e$^-$ $\rightarrow$ I + 6F	3.8	−9.58	111.33		106.07
IF$_6^-$ (C_{3v}) $\rightarrow$ I + 6F + e$^-$	1.8	−9.58	461.60		458.78
IF$_7$ $\rightarrow$ I + 7F	3.6	−9.97	386.53		380.97
IF$_8^-$ $\rightarrow$ I + 8F + e$^-$	3.0	−10.36	533.23		526.97

at 298 K for the former and −226.4 ± 0.6 kcal/mol at 0 K and −229.7 ± 0.6 kcal/mol at 298 K for the latter [75]. Good agreement for IF$_5$ is found for the extrapolated CV CBS value [$\sum D_0$ (0K) (DTQ)$_{CV}$] with the SO correction (see below) giving −202.6 kcal/mol at 298 K. The valence-only value for the Q5 extrapolation for IF$_5$ is also in reasonable agreement with experiment, but the extrapolated DTQ valence value is too negative. Excellent agreement is also found for IF$_7$ with a calculated value of −229.0 kcal/mol at 298 K at the (DTQ)$_{CV}$ level. For IF$_7$, the extrapolated DTQ valence value is too negative by ∼5 kcal/mol.

2.2. Calculating ΔE_{CV}

Most electronic structure calculations invoke the frozen core approximation in which the energetically lower lying orbitals, e.g. the 1s in fluorine, are excluded from the correlation treatment. In order to achieve thermo-chemical results within ±1 kcal/mol of experiment, it is necessary to account for CV (e.g. intershell $1s^2 - 2s^2 2p^5$ in F) correlation energy effects. CV calculations can be carried out with the weighted CV basis sets, i.e. cc-pwCVnZ, or their diffuse function augmented counterparts, aug-cc-pwCVnZ [76]. The CV correction ΔE_{CV} is then taken as the difference in energy between the valence electron correlation calculation and that with the appropriate core electrons included using the *same* basis sets. One must be careful to have the appropriate basis functions to correlate the core electrons. For example, for I, the cc-pwCVTZ-PP basis set contains functions up through g-functions to provide a consistent degree of angular correlation for the active 4d electrons. A CV calculations for I would involve all 25 electrons outside the RECP core, i.e. $4s^2$, $4p^6$, $5s^2$, $4d^{10}$ and $5p^5$. In general, CV effects tend to increase the TAE, but this is not always the case. As discussed above, this approach for ΔE_{CV}, which is taken from a large amount of work on first- and second-row compounds, works quite well, subject to issues with the valence basis sets as noted above.

2.3. Calculating ΔE_{SR}

Relativistic effects need to be included to obtain chemical accuracy, and up to three adjustments to the TAE are necessary in order to account for relativistic effects in atoms and molecules. The ΔE_{SR} correction to the atomization energy accounts for molecular scalar relativistic effects. For molecules containing first- or second-row atoms, ΔE_{SR} can be calculated using the spin-free, one-electron Douglas–Kroll–Hess (DKH) Hamiltonian [77–79]. ΔE_{SR} is defined as the difference in the atomization energy between the results obtained from basis sets recontracted for DKH calculations[52,80] and the atomization energy obtained with the normal valence basis set of the same quality. DKH calculations are usually carried out at the CCSD(T)/cc-pVTZ and the CCSD(T)/cc-pVTZ-DK levels of theory.

The use of RECPs accounts for the scalar relativistic effects on the atoms for which they are used. In our older work when ECPs were included, we evaluated ΔE_{SR} by using the expectation values for the two dominant terms in the Breit–Pauli Hamiltonian, the so-called mass–velocity and one-electron Darwin (MVD) corrections from configuration interaction singles and doubles (CISD) calculations with at least a triple-ζ basis set. The CISD(MVD) approach generally yields ΔE_{SR} values in good agreement (± 0.3 kcal/mol) with more accurate values from DKH calculations for most molecules. We do not recommend this approach anymore with the advent of DKH implementations in many codes and the broad availability of the DKH basis sets.

In order to estimate the errors from using the RECPs, CCSD(T) calculations can be performed at the second- or third-order DKH level with all-electron DKH basis sets [80–82]. For third-row transition metals, additional high angular momentum functions ($2f2g1h$) for correlating the metal 4f orbitals in CV calculations need to be included as these orbitals can be higher in energy than the metal 5s and 5p orbitals. One may have to rotate orbitals to get the proper occupied orbitals correlated, and this can be tricky for high symmetry molecules. The errors in the RECPS and the scalar relativistic correction for atoms with no RECP can be calculated as the difference in the electronic energies at the CCSD(T)-DK/awCVnZ-DK level ($\Delta E_{awCVnZ-DK}$) and at the CCSD(T) level with the RECP basis sets as $\Delta E_{PP,corr}$

$$\Delta E_{corr}(CV) = \Delta E_{awCVnZ-DK} - \Delta E_{awCVnZ-PP} \qquad (1.9)$$

or if only the valence electrons are included

$$\Delta E_{corr}(V) = \Delta E_{aVnZ-DK} - \Delta E_{aVnZ-PP}. \qquad (1.10)$$

2.4. Calculating ΔE_{SO}

It is necessary to include an SO relativistic correction which lowers the sum of the atomic energies (decreasing TAE) by replacing energies that correspond to an average over the available spin multiplets with energies for the lowest multiplets, as most electronic structure codes are only capable of producing spin multiplet averaged wave functions. Although the atomic SO correction, ΔE_{SO}, for first-row atoms are not too large (for F it is 0.39 kcal/mol), the atomic SO correction for heavier atoms can be much larger (for I it is 7.24 kcal/mol). These values are available from Moore's tables of experimental term energies for most atoms [83]. Obviously, such corrections are not negligible in considering accuracies of ±1 kcal/mol. It is usually assumed that molecular SO interactions are negligible for closed shell molecules or can be obtained from experiment [84,85]. The usual expression for calculating the SO expression is $\Sigma_J[(2J + 1)E_J]/\Sigma_J(2J + 1)$. Use of this approximation assumes that the highest term energy of the SO multiplet in the ground state is significantly below the lowest SO state of the excited state. However, this is not always true for transition metal atoms and can lead to issues with how to accurately include SO effects.

It may be necessary, however, to include SO corrections even for closed shell molecules containing heavy atoms such as I or Xe. For these molecules, second-order SO corrections may not be small when compared with chemical accuracy [50]. The lowest SO coupled eigenstates can be obtained by diagonalizing small SO matrices in a basis of pure spin ($\Lambda-S$) eigenstates. In each case, the electronic states used as an expansion basis can be restricted to all states (singlets and triplets) that correlate in the dissociation limit to ground state atomic products. For IF, this corresponds to 12 states (6 singlets and 6 triplets) and for IF_2^-, 18 states (9 singlets and 9 triplets) correlating to ground state products. The electronic states and SO matrix elements can be obtained in singles–only multireference configuration interaction calculations with a full valence complete active space (CAS) reference function. Again, at least a triple-ζ basis set should be used. Fully relativistic Dirac–Hartree–Fock CCSD(T) benchmark calculations can be performed to validate that the use of RECPs or other approaches to treat scalar relativistic effects in combination with separate SO calculations are properly accounting for relativistic corrections [86]. These calculations can be done at the CCSD(T) level for the valence orbitals with appropriate basis sets[87,88] and compared to the CCSD(T) RECP results.

Molecular SO corrections can also be calculated using the SO-DFT (density functional theory—SO) method [89], for example at the B3LYP level with an aVTZ-PP SO basis set on the heavy element and an aVTZ basis set on the lighter main group elements. It is recommended to de-contract these basis sets, however, since they do not generally have sufficient flexibility to describe SO effects with these methods. One can also calculate SO effects at the BLYP/TZ2P level using the SO approach and the scalar two–component zero–order regular approximation (ZORA)[90–94] as implemented in the ADF program [95,96]. The SO correction is taken as the difference between the ZORA and ZORA + SO values of a specific property. An issue with using ADF is that one only obtains the lowest SO state and that may not be the one of interest.

Why does one have to be concerned about SO effects? We provide three examples from our recent work to show the importance of these quantities. Table 1.1 clearly shows that the second-order SO effect for the closed shell IF_x molecules and ions cannot be neglected for chemical accuracy [73]. Table 1.2 shows the electron affinities of the fifth-row transition metal hexafluorides [97]. The electron affinities given in the second column were calculated without including SO effects and do not give the order expected from experimental reactivity measurements. When SO effects are included, the electron affinities, which are a direct measure for the oxidizer strength, increase monotonically from WF_6 to AuF_6 as do the experimental values. The inclusion of SO corrections is necessary to obtain even the correct qualitative order for the electron affinities. Without SO, the electron affinities of the MF_6 compounds follow the behavior expected for filling the t_{2g} orbitals (d_{xy}, d_{xz}, d_{yz}) as the atomic number increases from W to Au. However, this is not correct as the MF_6 molecules have differing electron configurations in the neutral and the anion. The largest change occurs for M = Os and M = Ir. OsF_6 has a d^2 occupancy with a $^3A_{1g}/D_{4h}$ ground state

Table 1.2 Electron affinities for MF_6 compounds, M a fifth-row transition metal [97]

Molecule	EA, kcal/mol	SO, kcal/mol	EA + SO, kcal/mol	EA + SO, eV	Experiment, eV	Reactivity Estimate, eV
W	70.9	1.90	72.8	3.16	3.50 ± 0.1, 3.36 (+0.04/−0.22)	>3.0
Re	104.7	0.88	105.6	4.58	>3.8	>3.90
Os	138.8	−2.30	136.5	5.92	5.93 ± 0.28	>4.7
Ir	132.2	5.98	138.2	5.99	6.50 ± 0.38	>5.46
Pt	163.0	0.62	163.6	7.09	7.00 ± 0.35	>6.76
Au	193.1	−3.90	189.2	8.20		>7.6

and the anion has a d^3 occupancy and a $^4A_{1g}/O_h$ ground state. SO corrections favor the ground state and hence the electron affinity decreases. In IrF_6, the opposite is predicted as the ground state is $^4A_{1g}/O_h$ from a d^3 occupancy and the anion has a $^3A_{1g}/D_{4h}$ ground state from a d^4 occupancy. The effect of SO is to change the difference by 0.36 eV (8.3 kcal/mol).

The final example showing the need for SO effects comes in explaining the potential energy surfaces of reactions of O atoms with halogen molecules, which are relevant to atmospheric chemistry [98]. Crossed molecular beam reactive scattering of ground-state (^3P) oxygen atoms with diatomic halogens XY shows that complexes can be formed and that the lifetime of the complex is related to the initial translational energy distribution [99–103]. The calculated values for the SO contributions for the OIX molecules with X = Cl, Br, and I are given in Table 1.3. We show values calculated using only the atomic contributions and those when the molecular contributions are also included. The results clearly show that the molecular contributions are substantial in decreasing the overall effect of SO and can be very large as shown for triplet IIO. As shown in Fig. 1.1, neglect of the molecular contribution would have left ^3OII above the product asymptote, which is not consistent with the crossed molecular beam scattering results. These SO corrections were calculated with the DFT SO options in NWChem because ^3IIO is not the lowest energy state with this geometry and a number of codes only provide the SO corrections to the lowest state for a given geometry.

2.5. Calculating ΔE_{HO}

Although CCSD(T) is a powerful tool for describing chemical phenomena, experience has shown that it is sometimes necessary to resort to levels of

Table 1.3 SO contributions to CCSD(T) atomization energies in kilocalories per mole [98].

Molecule	ΔE_{SO} atomic	ΔE_{SO}	ΣD_0 (0 K)
IOI $(C_{2v} - {}^1A_1)$	−14.70	−11.47	82.90
IIO $(C_s - {}^1A')$	−14.70	−12.21	71.03
IIO $(C_{\infty v} - {}^3\Sigma^1A')$	−8.30	−6.30	89.86
ClIO $(C_s - {}^1A')$	−8.30	−6.61	89.84
ClIO $(C_s - {}^3A'')$	−8.30	−6.94	73.04
IClO $(C_s - {}^1A')$	−8.30	−6.14	69.37
BrOI $(C_s - {}^1A')$	−10.96	−8.41	84.60
BrIO $(C_s - {}^1A')$	−10.96	−8.85	80.77
BrIO $(C_s - {}^3A'')$	−10.96	−9.08	64.62
IBrO $(C_s - {}^1A')$	−10.96	−8.62	64.02

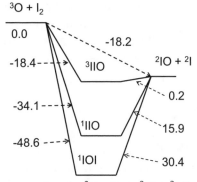

Figure 1.1 Potential energy surface for ${}^3O + I_2 \rightarrow {}^2IO + {}^2I$. Energies are enthalpies in kilocalories per mole, 298 K [98]. Energies are for a path proceeding from left to right.

theory capable of recovering an even larger fraction of the correlation energy. In the coupled cluster sequence of methods, the next more sophisticated treatments are CCSDT, which replaces the perturbative recovery of triple excitations in CCSD(T) with a fully iterative treatment, and CCSDTQ, which scales approximately as the N^{10}, where N is the number of basis functions. Both of these higher order correlation treatments are very expensive and therefore limited in the size of the molecules and basis sets with which they can be used. Because the CCSDT and CCSDTQ corrections to the atomization energy are often roughly the same size but of opposite signs, it is important to use a balanced treatment. We often try to match a cc-pV$(n + 1)Z$ basis set for the CCSDT calculation with a cc-pVnZ basis set for the CCSDTQ component, in recognition of the faster basis set convergence of the contributions of connected quadruples in the latter. The size of the higher order correction on frozen core calculations depends in part on the degree of multiconfigurational character and the size of the molecule. It can easily exceed our target accuracy of ± 1 kcal/mol. For example, the higher order correction in O_3 has been reported to be 3.2 kcal/mol [104]. Even for molecules whose wave functions are dominated by the HF configuration, higher order corrections become significant with more than four to five first-row atoms [105]. It is now possible to benchmark the energies of small molecules such as diatomics with full configuration interaction calculations for basis sets as large cc-pVTZ [106].

The remaining difference between CCSDTQ and full configuration CI (FCI), which represents the exact solution of the electronic, Born–Oppenheimer, molecular Schrödinger equation for a given basis set, is also estimated. For this purpose, we use a continued fraction approximant originally proposed by Goodson with SCF, CCSD and CCSD(T) energies [107]. Because that sequence of three energies proved to be unreliable, we replaced it with the CCSD, CCSDT and CCSDTQ energies. In the majority of cases (42 out of 49) where we have compared it to FCI for atomization energies, it has been found to improve the raw CCSDTQ results [104].

2.6. Calculating ΔE_{ZPE}

To convert vibrationless atomization energies, ΣD_e to ΣD_0 and ultimately to ΔH_f, we require as accurate vibrational frequencies as possible to calculate molecular zero-point energy vibrational corrections, ΔE_{ZPE}, and the temperature dependence of ΔE and ΔH to obtain these quantities at 298 K. For diatomics, one can calculate a potential energy curve to obtain anharmonic zero-point energies using a Dunham expansion[108] and then obtaining the harmonic and anharmonic components. For polyatomic molecules, one can calculate the harmonic frequencies as accurately as possible and average these values with the experimental fundamental frequencies to estimate the anharmonic zero-point energy [109]. One can also calculate the anharmonic corrections by using a second-order perturbative treatment using finite difference evaluations of third and semi-diagonal fourth derivatives, but this can become computationally expensive [110]. The largest correction to the zero-point energy is for the stretching modes involving light atoms, especially H atoms. One way to correct for the anharmonic components is to scale the A—H bonds to known harmonic frequencies from high-level calculations or from experiment. For example, one can scale $C-H$ stretches with respect to CH_4, $N-H$ stretches with respect to NH_3 and $O-H$ stretches with respect to H_2O. One can then use the scaled harmonic frequencies with the approximations given in the above references. The zero-point energy correction can be substantial and can contribute significantly to the calculation of the heats of formation for larger molecules. For example, the zero-point energy in $n\text{-}C_8H_{18}$ is estimated to be ~ 152 kcal/mol, and clearly one will have to make careful estimates of this value to achieve chemical accuracy [111].

2.7. Diagonal Born–Oppenheimer Correction

Another correction due to nuclear motion may need to be added for very high accuracy calculations [112]. For molecules containing hydrogen or other light elements, the first-order, adiabatic correction to the commonly employed Born–Oppenheimer approximation [113], in which electronic and nuclear motions are separated, can introduce significant errors to atomization energies and bond lengths, if very high-accuracy is desired. For instance, in the case of CH_2 ($^1A_1$), this effect reduces the atomization energy by -0.15 kcal/mol. In most of our recent work, we have evaluated this component at the CCSD(FC)/aug-cc-pVTZ level of theory.

2.8. Calculating ΔH_f

The calculation of ΔH_f at 0 K and 298 K requires atomic information from experiment. The first choice that one must make is which heat of formation of the atom to choose. For the heats of formation of the main group elements with relatively low Z, this is not inherently a difficult choice given that there are a number of databases available. The issue is with the accuracy of the data in the databases [75,114–116]. For example, the value for $\Delta H_f^0(B)$ has changed over time. The original JANAF value [75] was $\Delta H_f^0(0\,K, B) = 132.7 \pm 2.9$ kcal/mol. Storms and Mueller [117] recommended a much larger value of 136.2 ± 0.2 kcal/mol which was further analyzed by Ruscic and co-workers [118]. Martin and Taylor [119] calculated the atomization energies of BF and BF_3 using a composite approach based on CCSD(T), used these results in an analysis of the heat of formation of the boron atom, and came to a similar conclusion as that of Ruscic [118]. More recently, Karton and Martin [120] revised the heat of formation of the B atom to 135.1 ± 0.2 kcal/mol on the basis of the experimental heats of formation of BF_3 [75] and B_2H_6 coupled with W4 calculations of their total atomization energies. This is the current recommended value. We have recently used a similar approach to revise the heat of formation of the As atom in the gas phase [121]. A set of useful thermodynamic data tables with highly accurate atomic heats of formation for a few atoms are those developed by Ruscic and co-workers in the Active Thermochemical Tables (ATcT) project [122–124].

The whole question of transition metal heats of formation is more complicated. For example, the JANAF tables do not list many metal atom heats of formation. Not only are the error bars usually beyond chemical accuracy for a single-metal atom (usually ±1.5 to ±2.0 kcal/mol or larger) the actual values show differences larger than the error bars in different sources, and error bars are often not given. In addition, the issue of SO splitting corrections raised above needs to be carefully considered. For example, the ground states of Cr and Mo are ^7S, but the ground state of W is a ^3D with the ^7S state 8.44 kcal/mol higher in energy. Thus, it makes sense to calculate the dissociation energy with respect to the ^7S state of W, which is easy to calculate and then correct it with the experimental atomic excitation energy.

The standard heats of formation of compounds at 298 K can be calculated by using standard thermodynamic and statistical mechanics expressions in the rigid rotor–harmonic oscillator approximation [125] and the appropriate corrections for the heat of formation of the atoms [126]. In this approach, the temperature dependence can contain errors, especially if internal rotors are involved or highly anharmonic, coupled low-frequency vibrational modes. This approach is an approximation to the prediction of the temperature dependence. In addition, for larger molecules, one must be concerned about the potential for other low-energy conformers, which may contribute to the heat of formation. As the chemical system and the size of the ZPE both increase, it becomes increasingly difficult to prevent the uncertainty associated with the vibrational energy calculation from simultaneously growing. It is important to remember that the temperature correction for the heat of formation of an atom is based on the thermal correction for the element in its most stable state.

What is the minimal level that one can use to evaluate Eqn (1.1) with some hope of attaining chemical accuracy of ±1 kcal/mol? We note that RECPs should be used as appropriate in the following description. The first step is to calculate a geometry at the highest possible level. For the approach outlined now, one would optimize the geometries at the CCSD(T)/aug-cc-pVDZ and CCSD(T)/aug-cc-pVTZ levels and use the latter geometry in a single-point CCSD(T)/aug-cc-pVQZ calculation. If such geometries are too expensive to calculate, one can use an MP2/aug-cc-pVTZ optimized geometry or a geometry calculated at the density functional theory level with an appropriately chosen exchange–correlation functional. The first term ΔE_{elec}(CBS) should be evaluated up through CCSD(T)/aug-cc-pVQZ and extrapolated with the D and T energies to the CBS limit using Eqn (1.5). The second term ΔE_{CV} should be calculated at least at the CCSD(T)/aug-cc-pwCVTZ level. The ΔE_{SR} terms can be evaluated as the difference between the CCSD(T)/cc-pVTZ and the CCSD(T)/cc-pVTZ-DK calculations. The atomic SO corrections for ΔE_{SO} should be calculated from Moore's Tables and molecular ones obtained at the SO density functional theory level. The term ΔE_{HO} is usually ignored. The term ΔE_{ZPE} can be calculated at the highest possible affordable level and then appropriate averaging and scaling is done on a per molecule basis as described above. This is not a formal prescription but is one that we have found to be useful in predicting thermodynamic properties of many compounds.

2.9. Computational Codes

Most of the types of calculations described above can be performed with computer codes like MOLPRO [127], NWChem [128,129], Gaussian [130], or MOLFDIR. [131].

3. HOW ACCURATE AND RELIABLE IS THE FPD METHOD?

Much of the theoretical data associated with the large number of calculations performed with the FPD approach have been stored in the Computational Results Database[53,132,133] together with the available experimental information on the same systems. The most recent version of the database contains more than 115,000 entries covering 381 molecules and 41 atoms. An "entry" is defined as a set optimized structural parameters, a set of vibrational frequencies, an atomization energy, an ionization potential, an electron affinity, etc. While the scope of the database is rather broad, with entries spanning a very wide range of methods and basis sets, the emphasis is on high accuracy. Most of the high-accuracy results have appeared in print. The existence of the database facilitates statistical analyses of the performance of the FPD and other model chemistries.

Statistical comparisons of the FPD approach for atomization energies using experimental data in the CRDB with reported uncertainties of ± 2 kcal/mol or better can provide insight into its likely accuracy and reliability. At the high end of the spectrum of computational methods described throughout Section 1.2 of this report, with the largest available basis sets and corrections for higher order correlation effects, we find a mean absolute deviation (ε_{MAD}) of 0.19 kcal/mol and a root-mean-square deviation (ε_{RMS}) of 0.31 kcal/mol (139 comparisons). An average of the five CBS energies produced by Eqns (1.4)–(1.8) was used to estimate the basis set limit. Restricting the comparison to only those systems with experimental uncertainties ≤ 0.2 kcal/mol, ε_{MAD} drops to 0.09 kcal/mol and ε_{RMS} drops to 0.12 kcal/mol (75 comparisons). With slightly less demanding computational procedures employing valence basis sets no larger than aug-cc-pVQZ (or aug-cc-pV$(Q + d)Z$ for second-row compounds), CV corrections limited to CCSD(T)(CV)/cc-pwCVTZ, scalar relativistic corrections obtained from CCSD(T)-DKH/cc-pVTZ-DK calculations and no consideration of higher order effects, we find $\varepsilon_{MAD} = 0.49$ kcal/mol and $\varepsilon_{RMS} = 0.78$ kcal/mol. At the low end, the largest errors exceed 3 kcal/mol compared to the high end where they barely exceed 1 kcal/mol. For this low-end approach, a single CBS extrapolation formula, the mixed Gaussian/exponential expression in Eqn (1.5), was used. This relatively good showing for such comparatively modest calculations is partially due to fortuitous cancellation of errors. To put these results in perspective, the raw CCSD(T)/aug-cc-pVQZ level of theory yields $\varepsilon_{MAD} = 4.98$ kcal/mol and $\varepsilon_{RMS} = 6.30$ kcal/mol.

ACKNOWLEDGMENTS

This work was supported in part by the Chemical Sciences, Geosciences and Biosciences Division, Office of Basic Energy Sciences, U.S. Department of Energy (DOE) under grant no. DE-FG02-03ER15481 (catalysis center program). DAD also thanks the Robert Ramsay Chair Fund of the University of Alabama and Argonne National Laboratory for support.

REFERENCES

[1] Bartlett, R. J.; Musial, M. Coupled-Cluster Theory in Quantum Chemistry; *Rev. Mod. Phys.* **2007**, *79*, 291–352.
[2] Dunning, T. H., Jr. Gaussian Basis Sets for Use in Correlated Molecular Calculations. I. The Atoms Boron through Neon and Hydrogen; *J. Chem. Phys.* **1989**, *90*, 1007–1023.
[3] Peterson, K. A. Gaussian Basis Sets Exhibiting Systematic Convergence to the Complete Basis Set Limit In. *Annual Reports in Computational Chemistry*; Spellmeyer, D. C., Wheeler, R. A., Eds.; Elsevier: 2007; Vol. 3; Chapter 11, pp 195–206.

[4] Purvis, G. D., III.; Bartlett, R. J. A Full Coupled-Cluster Singles and Doubles Model: The Inclusion of Disconnected Triples; *J. Chem. Phys.* **1982**, *76*, 1910–1918.

[5] Raghavachari, K.; Trucks, G. W.; Pople, J. A.; Head-Gordon, M. A Fifth-Order Perturbation Comparison of Electron Correlation Theories; *Chem. Phys. Lett.* **1989**, *157*, 479–483.

[6] Watts, J. D.; Gauss, J.; Bartlett, R. J. Coupled-Cluster Methods with Noniterative Triple Excitations for Restricted Open-Shell Hartree–Fock and Other General Single Determinant Reference Functions. Energies and Analytical Gradients; *J. Chem. Phys.* **1993**, *98*, 8718–8733.

[7] Feller, D.; Peterson, K. A.; Grant Hill, J. On the Effectiveness of CCSD(T) Complete Basis Set Extrapolations for Atomization Energies; *J. Chem. Phys.* **2011**, *135*, 044102 (18 pages).

[8] Peterson, K. A.; Feller, D.; Dixon, D. A. Chemical Accuracy in Ab Initio Thermochemistry and Spectroscopy: Current Strategies and Future challenges; *Theor. Chem. Acc.* **2012**, *131*, 1079 (20 pages).

[9] Feller, D.; Peterson, K. A.; Dixon D. A. Further benchmarks of a composite, convergent, statistically-calibrated coupled cluster approach for thermochemical and spectroscopic studies, *Mol. Phys.* published online, May 22, 2012.

[10] Dixon, D. A.; Feller, D. F.; Peterson, K. A. Accurate Calculations of the Electron Affinity and Ionization Potential of the Methyl Radical; *J. Phys. Chem. A* **1997**, *101*, 9405–9409.

[11] Feller, D.; Dixon, D. A.; Peterson, K. A. Heats of Formation of Simple Boron Compounds; *J. Phys. Chem. A* **1998**, *102*, 7053–7059.

[12] Dixon, D. A.; Feller, D.; Peterson, K. A. Heats of Formation and Ionization Energies of NH_x, $x = 0–3$; *J. Chem. Phys.* **2001**, *115*, 2576–2581.

[13] Feller, D.; Dixon, D. A. Extended Benchmark Studies of Coupled Cluster Theory through Triple Excitations; *J. Chem. Phys.* **2001**, *115*, 3484–3496.

[14] Ruscic, B.; Wagner, A. F.; Harding, L. B.; Asher, R. L.; Feller, D.; Dixon, D. A.; Peterson, K. A.; Song, Y.; Qian, X.; Ng, C.-Y.; Liu, J.; Chen, W.; Schwenke, D. W. On the Enthalpy of Formation of Hydroxyl Radical and Gas-Phase Bond Dissociation Energies of Water and Hydroxyl; *J. Phys. Chem. A* **2002**, *106*, 2727–2747.

[15] Pople, J. A.; Head-Gordon, M.; Fox, D. J.; Raghavachari, K.; Curtiss, L. A. Gaussian-1 Theory: A General Procedure for Prediction of Molecular Energies; *J. Chem. Phys.* **1989**, *90*, 5622–5629.

[16] Curtiss, L. A.; Jones, C.; Trucks, G. W.; Raghavachari, K.; Pople, J. A. Gaussian-1 Theory of Molecular Energies for 2nd Row Compounds; *J. Chem. Phys.* **1990**, *93*, 2537–2545.

[17] Curtiss, L. A.; Raghavachari, K.; Trucks, G. W.; Pople, J. A. Gaussian-2 Theory for Molecular Energies of First- and Second-Row Compounds; *J. Chem. Phys.* **1991**, *94*, 7221–7230.

[18] Curtiss, L. A.; Raghavachari, K.; Redfern, P. C.; Pople, J. A. Gaussian-3 (G3) Theory for Molecules Containing First and Second-Row Atoms; *J. Chem. Phys.* **1998**, *109*, 7764–7776.

[19] Curtiss, L. A.; Redfern, P. C.; Raghavachari, K. Gaussian-4 Theory; *J. Chem. Phys.* **2007**, *126*, 084108 (12 pages).

[20] Curtiss, L. A.; Redfern, P. C.; Raghavachari, K. Gaussian-4 Theory Using Reduced Order Perturbation Theory; *J. Chem. Phys.* **2007**, *127*, 084108 (8 pages).

[21] Nyden, Marc. R.; Petersson, G. A. Complete Basis Set Correlation Energies. I. The Asymptotic Convergence of Pair Natural Orbital Expansions; *J. Chem. Phys.* **1981**, *75*, 1843–1856.

[22] Montgomery, J. A., Jr.; Ochterski, J. W.; Petersson, G. A. A Complete Basis Set Model Chemistry. IV. An Improved Atomic Pair Natural Orbital Method; *J. Chem. Phys.* **1994**, *101*, 5900–5909.

[23] Petersson, G. A. Complete Basis Set Thermochemistry and Kinetics, Chapter 13 In *Computational Thermochemistry*; Irikura, Karl K, Frurip, David J., Eds.; *ACS Symposium Series No. 677*; 1998; pp 237–266 Washington, D. C.

[24] Montgomery, J. A., Jr.; Frisch, M. J.; Ochterski, J. W.; Petersson, G. A. A Complete Basis Set Model Chemistry. VII. Use of the Minimum Population Localization Method; *J. Chem. Phys.* **2000**, *112*, 6532–6542.

[25] Zhong, S.; Barnes, E. C.; Petersson, George A. Uniformly Convergent *n*-tuple-z Augmented Polarized (*n*ZaP) Basis Sets for Complete Basis Set Extrapolations. I. Self-consistent Field Energies; *J. Chem. Phys.* **2008**, *129*, 184116 (12 pages).

[26] Martin, J. M. L.; Oliveira, G. d. Towards Standard Methods for Benchmark Quality Ab Initio Thermochemistry – W1 and W2 Theory; *J. Chem. Phys.* **1999**, *111*, 1843–1856.

[27] Boese, A. D.; Oren, M.; Atasoylu, O.; Martin, J. M. L.; Kallay, M.; Gauss, J. W3 Theory: Robust Computational Thermochemistry in the kJ/mol Accuracy Range; *J. Chem. Phys.* **2004**, *120*, 4129–4141.

[28] Karton, A.; Rabinovich, E.; Martin, J. M. L.; Ruscic, B. W4 Theory for Computational Thermochemistry: In Pursuit of Confident sub-kJ/mol Predictions; *J. Chem. Phys.* **2006**, *125*, 144108 (17 pages).

[29] Karton, A.; Taylor, P. R.; Martin, J. M. L. Basis Set Convergence of Post-CCSD Contributions to Molecular Atomization Energies; *J. Chem. Phys.* **2007**, *127*, 064104 (11 pages).

[30] Tajti, A.; Szalay, P. G.; Császár, A. G.; Kállay, M.; Gauss, J.; Valeev, E. F.; Flowers, B. A.; Vázquez, J.; Stanton, J. F. HEAT: High Accuracy Extrapolated Ab Initio Thermochemistry; *J. Chem. Phys.* **2004**, *121*, 11599–11613.

[31] Bomble, Y. J.; Vázquez, J.; Kállay, M.; Michauk, C.; Szalay, P. G.; Császár, A. G.; Gauss, J.; Stanton, J. F. High-Accuracy Extrapolated Ab Initio Thermochemistry. II. Minor Improvements to the Protocol and a Vital Simplification; *J. Chem. Phys.* **2006**, *125*, 064108 (8 pages).

[32] Harding, M. E.; Vazquez, J.; Ruscic, B.; Wilson, A. K.; Gauss, J.; Stanton, J. F. High-Accuracy Extrapolated Ab Initio Thermochemistry. III. Additional Improvements and Overview; *J. Chem. Phys.* **2008**, *128*, 114111 (15 pages).

[33] DeYonker, N. J.; Cundari, T. R.; Wilson, A. K. The Correlation Consistent Composite Approach (ccCA): An Alternative to the Gaussian-*n* methods; *J. Chem. Phys.* **2006**, *124*, 114104 (17 pages).

[34] DeYonker, N. J.; Grimes, T.; Yokel, S.; Dinescu, A.; Mintz, B.; Cundari, T. R.; Wilson, A. K. The Correlation-Consistent Composite Approach: Application to the G3/99 Test Set; *J. Chem. Phys.* **2006**, *125*, 104111 (15 pages).

[35] DeYonker, N. J.; Peterson, K. A.; Steyl, G.; Wilson, A. K.; Cundari, T. R. Quantitative Computational Thermochemistry of Transition Metal Complexes; *J. Phys. Chem. A* **2007**, *111*, 11269–11277.

[36] DeYonker, N. J.; Williams, T. G.; Imel, A. E.; Cundari, T. R.; Wilson, A. K. Accurate Thermochemistry for Transition Metal Complexes from First-Principles Calculations; *J. Chem. Phys.* **2009**, *131*, 024106 (9 pages).

[37] Prascher, B. P.; Lai, J. D.; Wilson, A. K. The Resolution of the Identity Approximation Applied to the Correlation Consistent Composite Approach; *J. Chem. Phys.* **2009**, *131*, 044130 (12 pages).

[38] DeYonker, N. J.; Wilson, B. R.; Pierpont, A. W.; Cundari, T. R.; Wilson, A. K. Toward the Intrinsic Error of the Correlation Consistent Composite Approach (ccCA); *Mol. Phys.* **2009**, *107*, 1107–1121.

[39] Hill, J. G.; Mitrushchenkov, A.; Yousaf, K. E.; Peterson, K. A. Accurate Ab Initio Ro-vibronic Spectroscopy of the $^2\Pi$ CCN Radical using Explicitly Correlated Methods; *J. Chem. Phys.* **2011**, *135*, 144309 (12 pages).

[40] Knowles, P. J.; Hampel, C.; Werner, H. J. Coupled Cluster Theory for High Spin, Open Shell Reference Wave Functions; *J. Chem. Phys.* **1994**, *99*, 5219–5227.

[41] Lee, T. J.; Taylor, P. R. A Diagnostic for Determining the Quality of Single-Reference Electron Correlation Methods; *Int. J. Quantum Chem. Symp* **1989**, *23*, 199–207.

[42] Janssen, C. L.; Nielsen, I. M. B. New Diagnostics for Coupled-Cluster and Møller–Plesset Perturbation Theory; *Chem. Phys. Lett.* **1998**, *290*, 423–430.

[43] Lee, T. J. Comparison of the T_1 and D_1 Diagnostics for Electronic Structure Theory: A New Definition for the Open-shell D_1 Diagnostic; *Chem. Phys. Lett.* **2003**, *372*, 362–367.

[44] Kendall, R. A.; Dunning, T. H., Jr.; Harrison, R. J. Electron Affinities of the First-Row Atoms Revisited. Systematic Basis Sets and Wave Functions; *J. Chem. Phys.* **1992**, *96*, 6796–6806.

[45] Dunning, T. H., Jr.; Peterson, K. A.; Wilson, A. K. Gaussian Basis Sets for Use in Correlated Molecular Calculations. X. The Atoms Aluminum through Argon Revisited; *J. Chem. Phys.* **2001**, *114*, 9244–9253.

[46] *Methods in Computational Chemistry*; Wilson, S., Ed.Vol. 2; Plenum Press: New York, 1988.

[47] Hess, B. A.; Dolg, M. *Relativistic Quantum Chemistry with Pseudopotentials and Transformed Hamiltonians. Wiley Series in Theoretical Chemistry*; John Wiley & Sons: Chichester, 2002; Vol. 57.

[48] Reiher, M. Relativistic Douglas–Kroll–Hess Theory; *Wiley Interdisciplinary Reviews: Comp. Mol. Sci.* **2012**, *2*, 139–149.

[49] Küchle, W.; Dolg, M.; Stoll, H.; Preuss, H. Pseudopotentials of the Stuttgart/Dresden Group **1998** (Revision: Tue Aug 11, 1998). http://www.theochem.uni-stuttgart.de/pseudopotentiale.

[50] Feller, D.; Peterson, K. A.; de Jong, W. A.; Dixon, D. A. Performance of Coupled Cluster Theory in Thermochemical Calculations of Small Halogenated Compounds; *J. Chem. Phys.* **2003**, *118*, 3510–3522.

[51] Peterson, K. A.; Figgen, D.; Dolg, M.; Stoll, H. Energy-Consistent Relativistic Pseudopotentials and Correlation Consistent Basis Sets for the 4d Elements Y–Pd; *J. Chem. Phys.* **2007**, *126*, 124101 Peterson, K. A. unpublished basis sets for the first and third row transition metals.

[52] EMSL basis set library: https://bse.pnl.gov/bse/portal.

[53] Feller, D. The Role of Databases in Support of Computational Chemistry Calculations; *J. Comp. Chem.* **1996**, *17*, 1571–1586.

[54] Schuchardt, K. L.; Didier, B. T.; Elsethagen, T.; Sun, L.; Gurumoorthi, V.; Chase, J.; Li, J.; Windus, T. L. Basis Set Exchange: A Community Database for Computational Sciences; *J. Chem. Inf. Model.* **2007**, *47*, 1045–1052.

[55] http://tyr0.chem.wsu.edu/~kipeters/basis.html.

[56] Feller, D. Application of Systematic Sequences of Wave Functions to the Water Dimer; *J. Chem. Phys.* **1992**, *96*, 6104–6114.

[57] Feller, D. The Use of Systematic Sequences of Wave Functions for Estimating the Complete Basis Set, Full Configuration Interaction Limit in Water; *J. Chem. Phys.* **1993**, *98*, 7059–7071.

[58] Peterson, K. A.; Woon, D. E.; Dunning, T. H., Jr. Benchmark Calculations with Correlated Molecular Wave Functions. IV. The Classical Barrier Height of the H + H$_2$ → H$_2$ + H Reaction; *J. Chem. Phys.* **1994**, *100*, 7410–7415.

[59] Martin, J. M. L. Ab Initio Total Atomization Energies of Small Molecules — Towards the Basis Set Limit; *Chem. Phys. Lett.* **1996**, *259*, 669–678.

[60] Helgaker, T.; Jorgensen, P.; Klopper, W.; Koch, H.; Olsen, J.; Wilson, A. K. Basis-Set Convergence in Correlated Calculations on Ne, N_2, and H_2O; *Chem. Phys. Lett.* **1998**, *286*, 243–252.

[61] Schwenke, D. W. The Extrapolation of One-electron Basis Sets in Electronic Structure Calculations: How it Should Work and How it can be Made to Work; *J. Chem. Phys.* **2005**, *122*, 014107 (7 pages).

[62] Li, S.; Dixon, D. A. Molecular Structures and Energetic of the $(TiO_2)_n$ ($n = 1-4$) Clusters and Their Anions; *J. Phys. Chem. A* **2008**, *112*, 6646–6666.

[63] Li, S.; Hennigan, J. M.; Dixon, D. A.; Peterson, K. A. Accurate Thermochemistry for Transition Metal Oxide Clusters; *J. Phys. Chem. A* **2009**, *113*, 7861–7877.

[64] Li, S.; Dixon, D. A. Molecular Structures and Energetics of the $(ZrO_2)_n$ and $(HfO_2)_n$ ($n = 1-4$) Clusters and Their Anions; *J. Phys. Chem. A* **2010**, *114*, 2665–2683.

[65] Werner, H.-J.; Knizia, G.; Adler, T. B.; Marchetti, O. Benchmark Studies for Explicitly Correlated Perturbation- and Coupled Cluster Theories; *Z. Phys. Chem.* **2010**, *224*, 493–511.

[66] Klopper, W.; Manby, F. R.; Ten-no, S.; Valeev, E. F. R12 Methods in Explicitly Correlated Molecular Electronic Structure Theory; *Int. Rev. Phys. Chem.* **2006**, *25*, 427–468.

[67] Shiozaki, T.; Valeev, E. F.; Hirata, S. Explicitly Correlated Coupled-Cluster Methods In. *Annual Reports in Computational Chemistry*; Wheeler, R. A., Ed.; Elsevier: 2009; Vol. 5; Chapter 6, pp 131–148.

[68] Peterson, K. A.; Adler, T. B.; Werner, H.-J. Systematically Convergent Basis Sets for Explicitly Correlated Wavefunctions: The Atoms H, He, B–Ne, And Al–Ar; *J. Chem. Phys.* **2008**, *128*, 084102 (12 pages).

[69] Peterson, K. A.; Yousaf, K. E. Optimized Auxiliary Basis Sets for Explicitly Correlated Methods; *J. Chem. Phys.* **2010**, *133*, 174116 (8 pages).

[70] Knizia, G.; Adler, T. B.; Werner, H.-J. Simplified CCSD(T)-F12 Methods: Theory and Benchmarks; *J. Chem. Phys.* **2009**, *130*, 054104 (20 pages).

[71] Hill, J. G.; Peterson, K. A.; Knizia, G.; Werner, H.-J. Extrapolating MP2 and CCSD Explicitly Correlated Correlation Energies to the Complete Basis Set Limit with First and Second Row Correlation Consistent Basis Sets; *J. Chem. Phys.* **2009**, *131*, 194105 (13 pages).

[72] Feller, D.; Peterson, K. A.; Hill, J. G. Calibration Study of the CCSD(T)-F12a/b Methods for C2 and Small Hydrocarbons; *J. Chem. Phys.* **2010**, *133*, 184102 (17 pages).

[73] Dixon, D. A.; Grant, D. J.; Christe, K. O.; Peterson, K. A. The Structure and Heats of Formation of Iodine Fluorides and the Respective Closed Shell Ions from CCSD(T) Electronic Structure Calculations and Reliable Prediction of the Steric Activity of the Free Valence Electron Pair in ClF_6^-, BrF_6^- and IF_6^-; *Inorg. Chem.* **2008**, *47*, 5485–5494.

[74] Grant, D. J.; Wang, T.-H.; Dixon, D. A.; Christe, K. O. Heats of Formation of XeF_3^+, XeF_3^-, XeF_5^+, XeF_7^+, XeF_7, and XeF_8 from High Level Electronic Structure Calculations; *Inorg. Chem.* **2010**, *49*, 261–270.

[75] Chase, M. W., Jr. NIST-JANAF Themochemical Tables J. Phys. Chem. Ref. Data, Monograph 9 In *NIST Chemistry WebBook, NIST Standard Reference Database Number 69, August 1996*, 4th ed.; Mallard, W. G., Ed.; National Institute of Standards and Technology: Gaithersburg MD, 1998; pp 1–1951; 20899; http://webbook.nist.gov/chemistry/.

[76] Peterson, K. A.; Dunning, T. H., Jr. Accurate Correlation Consistent Basis Sets for Molecular Core–Valence Correlation Effects. The Second Row Atoms Al–Ar, and the First Row Atoms B–Ne Revisited; *J. Chem. Phys.* **2002**, *117*, 10548–19560.

[77] Douglas, M.; Kroll, N. M. Quantum Electrodynamical Corrections to the Fine Structure of Helium; *Ann. Phys.* **1974**, *82*, 89–155.

[78] Hess, B. A. Applicability of the No-Pair Equation with Free-Particle Projection Operators to Atomic and Molecular Structure Calculations; *Phys. Rev. A* **1985**, *32*, 756–763.

[79] Hess, B. A. Relativistic Electronic-Structure Calculations Employing a Two-Component No-Pair Formalism with External-Field Projection Operators; *Phys. Rev. A* **1986**, *33*, 3742–3748.

[80] de Jong, W. A.; Harrison, R. J.; Dixon, D. A. Parallel Douglas–Kroll energy and Gradients in NWChem. Estimating Scalar Relativistic Effects using Douglas–Kroll Contracted Basis Sets; *J. Chem. Phys.* **2001**, *114*, 48–53.

[81] Balabanov, N. B.; Peterson, K. A. Systematically Convergent Basis Sets for Transition Metals. I. All-Electron Correlation Consistent Basis Sets for the 3d Elements Sc–Zn; *J. Chem. Phys.* **2005**, *123*, 064107 (15 pages).

[82] Balabanov, N. B.; Peterson, K. A. Basis Set Limit Electronic Excitation Energies, Ionization Potentials, and Electron Affinities for the 3d Transition Metal Atoms: Coupled Cluster and Multireference Methods; *J. Chem. Phys.* **2006**, *125*, 074110 (10 pages).

[83] Moore, C. E. *Atomic Energy Levels as Derived from the Analysis of Optical Spectra, Volume 1, 2, and 3, U.S. National Bureau of Standards Circular 467*; U.S. Department of Commerce, National Technical Information Service: Washington, DC, 1949; COM-72-50282.

[84] Herzberg, G. *Molecular Spectra and Molecular Structure I. Spectra of Diatomic Molecules*; Van Nostrand Reinhold Co., Inc.: New York, 1950.

[85] Huber, K. P.; Herzberg, G. *Constants of Diatomic Molecules. Molecular Spectra and Molecular Structure*; Van Nostrand: Princeton, 1979; Vol. IV.

[86] Dixon, D. A.; de Jong, W. A.; Peterson, K. A.; Christe, K.; Schrobilgen, G. J. The Heats of Formation of Xenon Fluorides and the Fluxionality of XeF_6 From High Level Electronic Structure Calculations; *J. Am. Chem. Soc.* **2005**, *127*, 8627–8634.

[87] Dyall, K. G. Relativistic and Nonrelativistic Finite Nucleus Optimized Triple-Zeta Basis Sets for the 4 p, 5 p and 6 p Elements; *Theor. Chem. Acc.* **2002**, *108*, 335–340.

[88] de Jong, W. A.; Styszynski, J.; Visscher, L.; Nieuwpoort, W. C. Relativistic and Correlation Effects on Molecular Properties: The Interhalogens ClF, BrF, BrCl, IF, ICl, and IBr; *J. Chem. Phys.* **1998**, *108*, 5177–5184.

[89] Hess, B. A.; Marian, C. M.; Wahlgren, U.; Gropen, O. A Mean-Field Spin–Orbit Method Applicable to Correlated Wavefunctions; *Chem. Phys. Lett.* **1996**, *251*, 365–371.

[90] van Lenthe, E.; Ehlers, A. E.; Baerends, E. J. Geometry Optimizations in the Zero Order Regular Approximation for Relativistic Effects; *J. Chem. Phys.* **1999**, *110*, 8943–8953.

[91] van Lenthe, E.; Baerends, E. J.; Snijders, J. G. Relativistic Regular Two-Component Hamiltonians; *J. Chem. Phys.* **1993**, *99*, 4597–4610.

[92] van Lenthe, E.; Snijders, J. G.; Baerends, E. J. The Zero-Order Regular Approximation for Relativistic Effects: The Effect of Spin–Orbit Coupling in Closed Shell Molecules; *J. Chem. Phys.* **1994**, *105*, 6505–6517.

[93] van Lenthe, E.; Baerends, E. J.; Snijders, J. G. Relativistic Total Energy using Regular Approximations; *J. Chem. Phys.* **1994**, *101*, 9783–9792.

[94] van Lenthe, E.; van Leeuwen, R.; Baerends, E. J.; Snijders, J. G. Relativistic Regular Two-Component Hamiltonians; *Int. J. Quantum Chem.* **1996**, *57*, 281–293.

[95] te Velde, G.; Bickelhaupt, F. M.; van Gisbergen, S. J. A.; Fonseca Guerra, C.; Baerends, E. J.; Snijders, J. G.; Ziegler, T. Chemistry with ADF; *J. Comp. Chem.* **2001**, *22*, 931–967 ADF2012, SCM, Theoretical Chemistry, Vrije Universiteit, Amsterdam, The Netherlands; http://www.scm.com.

[96] Fonseca Guerra, C.; Snijders, J. G.; te Velde, G.; Baerends, E. J. Towards an Order-N DFT Method; *Theor. Chem. Acc.* **1998**, *99*, 391–403.

[97] Craciun, R.; Picone, D.; Long, R. T.; Li, S.; Dixon, D. A.; Peterson, K. A.; Christe, K. O. Third Row Transition Metal Hexafluorides, Extraordinary Oxidizers and Lewis Acids: Electron Affinities, Fluoride Affinities, and Heats of Formation of WF_6, ReF_6, OsF_6, IrF_6, PtF_6, and AuF_6; *Inorg. Chem.* **2010**, *49*, 1056–1070.

[98] Grant, D. J.; Garner, E. B., III; Matus, M. H.; Nguyen, M. T.; Peterson, K. A.; Francisco, J. S.; Dixon, D. A. Thermodynamic Properties of the XO_2, X_2O, XYO, X_2O_2, and XYO_2 (X, Y = Cl, Br and I); *Isomers, J. Phys. Chem. A* **2010**, *114*, 4254–4265.

[99] Parrish, D. D.; Herschbach, D. R. Molecular Beam Chemistry: Persistent Collision Complex in Reaction of Oxygen Atoms with Bromine Molecules; *J. Am. Chem. Soc.* **1973**, *95*, 6133–6134.

[100] Dixon, D. A.; Parrish, D. D.; Herschbach, D. R. Possibility of Singlet-Triplet Transitions in Oxygen Exchange Reactions; *Faraday Discuss. Chem. Soc.* **1973**, *55*, 385–387.

[101] Grice, R. Reactive Scattering of Ground-State Oxygen Atoms; *Acc. Chem. Res.* **1981**, *14*, 37–42.

[102] Clough, P. N.; O'Neil, G. M.; Geddes, J. Crossed-beam Investigation of Translational Energy Effects in Oxygen Atom Reactions; *J. Chem. Phys.* **1978**, *69*, 3128–3135.

[103] Gorry, P. A.; Nowikow, C. V.; Grice, R. Reactive Scattering of a Supersonic Oxygen Atom Beam: O + Cl_2; *Molec. Phys.* **1979**, *37*, 347–359.

[104] Feller, D.; Peterson, K. A.; Dixon, D. A. A Survey of Factors Contributing to Accurate Theoretical Predictions of Atomization Energies and Molecular Structures; *J. Chem. Phys.* **2008**, *129*, 204105 (32 pages).

[105] Feller, D.; Dixon, D. A. Predicting the Heats of Formation of Model Hydrocarbons up to Benzene; *J. Phys. Chem. A* **2000**, *104*, 3048–3056.

[106] Gan, Z.; Grant, D. J.; Harrison, R. J.; Dixon, D. A. The Lowest Energy States of the Group IIIA – Group VA Heteronuclear Diatomics: BN, BP, AlN, and AlP from Full Configuration Interaction Calculations; *J. Chem. Phys.* **2006**, *125*, 124311 (6 pages).

[107] Goodson, D. Z. Extrapolating the Coupled-Cluster Sequence Toward the Full Configuration Interaction Limit; *J. Chem. Phys.* **2002**, *116*, 6948–6956.

[108] Dunham, J. L. The Energy Levels of a Rotating Vibrator; *Phys. Rev.* **1932**, *41*, 721–731.

[109] Grev, R. S.; Janssen, C. L.; Schaefer, H. F., III. Concerning Zero-Point Vibrational Energy Corrections to Electronic Energies; *J. Chem. Phys.* **1991**, *95*, 5128–5132.

[110] Califano, S. *Vibrational States*; Wiley: London, 1976.

[111] Pollack, L.; Windus, T. L.; de Jong, W. A.; Dixon, D. A. Thermodynamic Properties of the C5, C6, and C8 n-Alkanes from Ab Initio Electronic Structure Theory; *J. Phys. Chem. A* **2005**, *109*, 6934–6938.

[112] Valeev, E. F.; Scherrill, C. D. The diagonal Born–Oppenheimer Correction Beyond the Hartree–Fock Approximation; *J. Chem. Phys.* **2003**, *118*, 3921–3927.

[113] Born, M.; Oppenheimer, J. R. Zur Quantentheorie der Molekeln, On the Quantum Theory of Molecules; *Ann. Physik.* **1927**, *84*, 457–484.

[114] Gurvich, L. V.; Veyts, I. V.; Alcock, C. B.. Thermodynamic Properties of Individual Substances; Begell House: New York, 1996; Vol. 3.

[115] Wagman, D. D.; Evans, W. H.; Parker, V. B.; Schumm, R. H.; Halow, I.; Bailey, S. M.; Churney, K. L.; Nuttall, R. L. The NBS Tables of Chemical Thermodynamic Properties. Selected Values for Inorganic and C1 and C2 Organic Substances in SI Units; *J. Phys. Chem. Ref. Data* **1982**, *11* (Suppl. 2).

[116] Greenwood, N. N.; Earnshaw, A. *Chemistry of the Elements*; Pergamon Press: Oxford, 1984.

[117] Storms, E.; Mueller, B. Phase Relations and Thermodynamic Properties of Transition Metal Borides. I. The Molybdenum–Boron System and Elemental Boron; *J. Phys. Chem.* **1977**, *81*, 318–324.

[118] Ruscic, B.; Mayhew, C. A.; Berkowitz, J. Photoionization Studies of (BH3)n (n = 1,2); *J. Chem. Phys.* **1988**, *88*, 5580–5593.

[119] Martin, J. M. L.; Taylor, P. R. Revised Heat of Formation for Gaseous Boron: Basis Set Limit Ab Initio Binding Energies of BF3 and BF; *J. Phys. Chem. A* **1998**, *102*, 2995–2998.

[120] Karton, A.; Martin, J. M. L. Heats of Formation of Beryllium, Boron, Aluminum, and Silicon Re-examined by Means of W4 Theory; *J. Phys. Chem. A* **2007**, *111*, 5936–5944.

[121] Feller, D.; Vasiliu, M.; Grant, D. J.; Dixon, D. A. Thermodynamic Properties of Arsenic Compounds and the Heat of Formation of the As Atom from High Level Electronic Structure Calculations; *J. Phys. Chem. A* **2011**, *115*, 14667–14676.

[122] http://atct.anl.gov/.

[123] Ruscic, B.; Pinzon, R. E.; Morton, M. L.; von Laszevski, G.; Bittner, S. J.; Nijsure, S. G.; Amin, K. A.; Minkoff, M.; Wagner, A. F. Introduction to Active Thermochemical Tables: Several "Key" Enthalpies of Formation Revisited; *J. Phys. Chem. A* **2004**, *108*, 9979–9997.

[124] Ruscic, B.; Pinzon, R. E.; von Laszevski, G.; Kopdeboyina, D.; Burcat, A.; Leahy, D.; Montoya, D.; Wagner, A. F. Active Thermochemical Tables: Thermochemistry of the 21st Century; *J. Phys. Conf. Ser.* **2005**, *16*, 561–570.

[125] McQuarrie, D. A. *Statistical Mechanics*; University Science Books: Sausalito, CA, 2001.

[126] Curtiss, L. A.; Raghavachari, K.; Redfern, P. C.; Pople, J. A. Assessment of Gaussian-2 and Density Functional Theories for the Computation of Enthalpies of Formation; *J. Chem. Phys.* **1997**, *106*, 1063–1079.

[127] Werner, H. -J.; Knowles, P. J.; Knizia, G.; Manby, F. R.; Schütz, M.; Celani, P.; Korona, T.; Lindh, R.; Mitrushenkov, A.; Rauhut, G.; Shamasundar, K. R.; Adler, T. B.; Amos, R. D.; Bernhardsson, A.; Berning, A.; Cooper, D. L.; Deegan, M. J. O.; Dobbyn, A. J.; Eckert, F.; Goll, E.; Hampel, C.; Hesselmann, A.; Hetzer, G.; Hrenar, T.; Jansen, G.; Köppl, C.; Liu, Y.; Lloyd, A. W.; Mata, R. A.; May, A. J.; McNicholas, S. J.; Meyer, W.; Mura, M. E.; Nicklass, A.; O'Neill, D. P.; Palmieri, P.; Pflüger, K.; Pitzer, R.; Reiher, M.; Shiozaki, T.; Stoll, H.; Stone, A. J.; Tarroni, R.; Thorsteinsson, T.; Wang, M.; Wolf, A., *MOLPRO, Version 2010.1, A Package of Ab Initio Programs*, see http://www.molpro.net.

[128] Valiev, M.; Bylaska, E. J.; Govind, N.; Kowalski, K.; Straatsma, T. P.; van Dam, H. J. J.; Wang, D.; Nieplocha, J.; Apra, E.; Windus, T. L.; de Jong, W. A. NWChem: A Comprehensive and Scalable Open-Source Solution for Large Scale molecular simulations; *Comput. Phys. Commun.* **2010**, *181*, 1477–1489.

[129] Kendall, R. A.; Apra, E.; Bernholdt, D. E.; Bylaska, E. J.; Dupuis, M.; Fann, G. I.; Harrison, R. J.; Ju, J.; Nichols, J. A.; Nieplocha, J.; Straatsma, T. P.; Windus, T. L.; Wong, A. T. Computer Phys. Commun. **2000**, *128*, 260–283.

[130] Frisch, M. J.; Trucks, G. W.; Schlegel, H. B.; Scuseria, G. E.; Robb, M. A.; Cheeseman, J. R.; Scalmani, G.; Barone, V.; Mennucci, B.; Petersson, G. A.;

Nakatsuji, H.; Caricato, M.; Li, X.; Hratchian, H. P.; Izmaylov, A. F.; Bloino, J.; Zheng, G.; Sonnenberg, J. L.; Hada, M.; Ehara, M.; Toyota, K.; Fukuda, R.; Hasegawa, J.; Ishida, M.; Nakajima, T.; Honda, Y.; Kitao, O.; Nakai, H.; Vreven, T.; Montgomery, J. A., Jr.; Peralta, J. E.; Ogliaro, F.; Bearpark, M.; Heyd, J. J.; Brothers, E.; Kudin, K. N.; Staroverov, V. N.; Kobayashi, R.; Normand, J.; Raghavachari, K.; Rendell, A.; Burant, J. C.; Iyengar, S. S.; Tomasi, J.; Cossi, M.; Rega, N.; Millam, J. M.; Klene, M.; Knox, J. E.; Cross, J. B.; Bakken, V.; Adamo, C.; Jaramillo, J.; Gomperts, R.; Stratmann, R. E.; Yazyev, O.; Austin, A. J.; Cammi, R.; Pomelli, C.; Ochterski, J. W.; Martin, R. L.; Morokuma, K.; Zakrzewski, V. G.; Voth, G. A.; Salvador, P.; Dannenberg, J. J.; Dapprich, S.; Daniels, A. D.; Farkas, Ö.; Foresman, J. B.; Ortiz, J. V.; Cioslowski, J.; Fox, D. J. *Gaussian 09*; Gaussian, Inc: Wallingford CT, 2009.

[131] Pernpointer, M.; Visscher, L.; de Jong, W. A.; Broer, R. Parallelization of Four-Component Calculations. I. Integral Generation, SCF, and Four-Index Transformation in the Dirac–Fock Package MOLFDIR; *J. Comp. Chem.* **2000**, *21*, 1176–1186.

[132] Feller, D.; Peterson, K. A. An Examination of Intrinsic Error in Electronic Structure Methods Using the Environmental Molecular Sciences Laboratory Computational Results Database and the Gaussian-2 Set; *J. Chem. Phys.* **1998**, *108*, 154–176.

[133] Feller, D.; Peterson, K. A. A Re-examination of Atomization Energies for the Gaussian-2 Set of Molecules; *J. Chem. Phys.* **1999**, *110*, 8384–8396.

CHAPTER TWO

Ab Initio Composite Approaches: Potential Energy Surfaces and Excited Electronic States

Wanyi Jiang and Angela K. Wilson[1]

Department of Chemistry, Center for Advanced Scientific Computing and Modeling (CASCaM), University of North Texas, Denton, Texas, USA
[1]Corresponding author: E-mail: akwilson@unt.edu

Contents

Abstract

The correlation consistent composite approach (ccCA) is a thermochemical model devised to achieve energetics (i.e. enthalpies of formation, ionization potentials, and proton affinities) akin to those obtained using electronic correlation methods such as CCSD(T) at the complete basis set limit, but at much greater computational efficiency. While the approach has proven great utility, the prediction of features of potential energy surfaces, bond formation and bond breaking, and excited states can warrant a multireference wavefunction-based approach. This report illustrates the development of MR-ccCA, a multireference analog of ccCA, which is based upon a CASPT2 reference. The successful application of MR-ccCA has been demonstrated in the study of potential energy curves of C_2, spectroscopic constants of C_2, N_2, and O_2, and properties of silicon compounds and a few other first row and second row molecules.

Annual Reports in Computational Chemistry, Volume 8
ISSN 1574-1400,
http://dx.doi.org/10.1016/B978-0-444-59440-2.00002-8

1. INTRODUCTION

Computational chemistry has seen extensive applications in the theoretical prediction of energetic and chemical properties of molecules and materials. Coupled cluster (CC) theory, especially the CC method including singles, doubles and perturbative triples [CCSD(T)] [1] in combination with a large basis set has become the standard route for the prediction of accurate thermochemical and spectroscopic properties. However, CCSD(T) has an unfavorable scaling of computational cost that increases as N^7, where N increases with respect to the size of the molecular system. In order to reduce the overall computational cost, while retaining desirable accuracy, there have been many strategies that have been developed. Among the most widely used are composite approaches, which are designed to replicate the predictive abilities of a methodology such as CCSD(T)/large basis set, utilizing a series of less reliable, albeit, much more computationally cost efficient (in terms of computer time, memory, and disk space) method and basis set combinations.

Since 1989, composite approaches (a.k.a. model chemistries) [2–4] have advanced to effective methods that greatly enrich the predictive power of quantum chemistry. The enormous successes of composite approaches originate primarily from the additivity of different energetic components that include, but are not limited to, stationary geometrical parameters, vibrational frequencies, basis set effects, core–valence electron correlation, higher level valence electron correlation, scalar relativistic effects, spin–orbit coupling, and the non-Born–Oppenheimer correction, which can be treated separately and effectively using various theory/basis combinations. Essentially, composite approaches are optimal protocols of several theory/basis combinations in order to achieve a targeted accuracy at reduced cost. Efforts for such protocols were started by Pople and his co-workers in their proposition of Gaussian-1 [5,6]. The Gaussian-n approaches have evolved with the advances of new theories and basis sets and now includes multiple variants of G2 [7], G3 [8–10], and G4 [11,12]. However, the Gn methods include empirical parameters (i.e. higher-level correction, HLC) that are reliant on a test set of experimental data and may render unpredictable performance for molecular systems whose electronic structure is strikingly different from those in the initial test set. In contrast, the correlation consistent Composite Approach (ccCA) developed by our group [13–15] is parameter free and thus, in principle, gives consistent results for broad chemical problems; for example, ccCA predicts reaction barrier heights [16], which require a reliable description of transition states with strongly distorted geometries, by meeting a kinetically meaningful accuracy of $\pm 1.0\,\text{kcal}\,\text{mol}^{-1}$. Notable examples other than ccCA and Gn also include Wn [17–20], HEAT [21–23], the focal point method [24,25], CBS-n [26–33], CCSD(T)/CBS-based Feller–Peterson–Dixon method [4,34–40 and Chapter 1 of this volume], and multi-coefficient correlation method (MCCM) [41,42].

The successes of composite methods are widespread, and the methods have been routinely used in theoretical studies of accurate thermochemical properties. To provide several examples, ccCA [13–15] has predicted the thermochemical quantities, on average, within $\pm 1.0\,\text{kcal}\,\text{mol}^{-1}$ of the G3/05 set of 454 experimental data including dissociation energies, enthalpies of formation, ionization potentials, and proton and electron affinities; ccCA-TM predicted the enthalpy of formation for $Sc(C_5H_5)_3$ to be $20.3\,\text{kcal}\,\text{mol}^{-1}$ in excellent agreement with the experimental value of $20.0 \pm 1.4\,\text{kcal}\,\text{mol}^{-1}$; W4 [18,20] has reproduced total atomization energies (TAEs) of 35 molecules with an uncertainty of under $1.0\,\text{kJ}\,\text{mol}^{-1}$ ($0.24\,\text{kcal}\,\text{mol}^{-1}$) at a confidence interval of $\pm 3\sigma$ (99.7%).

Representative composite approaches such as W3, W4, and HEAT achieve a sub-kJ$\,\text{mol}^{-1}$ accuracy with respect to highly accurate experimental data such as the Active Thermochemical Tables (ATcT) [43–46] and the Computational Results Database (CRDB) [47], but at an extremely high computational cost, and thus their applications are limited to molecules of the smallest size, generally less than five nonhydrogen atoms. On the other hand, composite approaches such as G3, G4, and ccCA can attain an overall accuracy of $\pm 1.0\,\text{kcal}\,\text{mol}^{-1}$ relative to the established G3/05 set [48], which includes 270 enthalpies of formation, 105 ionization potentials, 63 electron affinities, 10 proton affinities of main group species, and 6 association energies of hydrogen bonded complexes, at a modest computational cost. Notably, ccCA is distinguishable from other approaches of comparable accuracy due to its absence of empirical parameterization, relatively low computational cost based on MP2, and remarkable performance for a wide spectrum of chemical problems. In a recent comparative study of various composite approaches, among the MP2- and MP4-based composite methods, ccCA is the closest to the W4 benchmark in terms of overall reliability and accuracy [20]. Recently, the applications of ccCA have also been successfully extended to heavy p-block elements [49], s-block metals

[50,51], and, with some modifications in formulation, to 3d (ccCA-TM [52–54]) and 4d transition metals (rp-ccCA [55]).

Similar to the majority of composite approaches, ccCA is based on single reference theories. Specifically, ccCA includes B3LYP for geometries and frequencies, HF for molecular orbitals, MP2 and CCSD(T) for electron correlation energies [15]. It is well known that single reference theories, particularly those of lower order such as MP2 and CCSD(T), may not be suitable for chemical systems with strong nondynamical correlation that require a multi-configuration or multireference method for a qualitatively correct description (see, Refs. [56–58]). The hierarchy of CC theory allows the inclusion of higher order excitations to progressively approach full configuration interaction (FCI), the exact solution to the Schrödinger equation. Consequently, high-order CC approaches may effectively recover the nondynamical correlation in multireference molecular systems. Martin et al. [20] have demonstrated that W4, which includes CCSDT, CCSDTQ, and CCSDTQ5, can predict thermodynamic data in excellent agreement with experimental data for molecules with significant nondynamical correlation. However, the highly accurate composite approaches (i.e. W4, W3, and HEAT) are based on CCSD(T), which, for example, fails to give even a qualitatively correct description of the potential energy curve of the ground state of N_2 at the bond-breaking region, requiring CCSDTQ56 for a qualitatively correct description [57]. Alternatively, modifications of CCSD(T) can be used to reduce the error originated from nondynamical correlation. For example, substitution of conventional CCSD(T) by CR-CC(2,3) [59] within ccCA has been examined for the description for bond-breaking processes [60].

There are many chemical situations for which single reference theories and their modifications may not give quantitatively correct or even qualitatively correct results. As such, a composite series of single reference theories may not eliminate the deficiency in recovering the nondynamical correlation in multireference molecular systems [61]. Consequently, multireference composite approaches with *genuine* multireference theories using a generic multi-configurational reference wavefunction have been sought for an accurate description of chemical systems with strong nondynamical correlation. Prior to our multireference analog of ccCA (MR-ccCA) [61–64], the systematic investigations of multireference variants of composite approach were sporadic and limited in nature. Multireference modifications of Gn (G2 and G3) theory [65], W1CAS, and W2CAS [66], and W2C-CAS-ACPF and W2C-CAS-AQCC [67] were proposed but were considered largely for the predictions of thermochemical properties at equilibrium geometries for main group species where single reference methods prevail.

Based upon ccCA, ideal features of a multireference composite approach include a complete basis set (CBS) extrapolated limit of a multireference theory upon which additive contributions can be included. Specifically, a relatively high-level theory with a basis set of a moderate size can be used to obtain valence electron correlation; the core electrons are correlated in one calculation to consider the interactions among core and valence electrons; scalar relativistic effects can be obtained by, for example, using the Douglas–Kroll–Hess Hamiltonian and the corresponding basis sets. Although a formal or explicit multireference composite approach that includes all essential energy components as mentioned above has seldom been utilized, until recently [68–82], historically individual energy components have long been tested and utilized for multireference methods such as icMRCI and CASPT2. For example, the CBS extrapolated limits of icMRCI and icMRCI+Q have been compared for the classical barrier height of the $H + H_2 \rightarrow H_2 + H$ reaction [83] and the classic dissociation energies of N_2 [84], using the correlation consistent basis set cc-pVxZ ($x = $ D, T, Q, 5) [85]. The core/valence interactions [86,87] and scalar relativistic effects were considered shortly after the advent of the respective correlation consistent basis sets [88–90]. Currently, CBS extrapolation and various additive corrections have often been used in recent studies of potential energy curves and properties of excited electronic states (see, Refs. 68–76). Shi et al. have considered a multireference composite approach based on icMRCI+Q, denoted as MRCI+Q/CV+DK+56, including core/valence corrections by cc-pCV5Z, scalar relativistic corrections by cc-pV5Z, and CBS extrapolations of cc-pV5Z and cc-pV6Z in the study of excited states of N_2 [68]. Smaller correlation consistent basis sets were also utilized in similar studies by Shi and co-workers [69–72]. Varandas has suggested a new CBS extrapolation scheme for multireference calculations [75–77]. As a more efficient alternative to icMRCI+Q, CASPT2 with the CBS extrapolated energy, core/valence correction, and scalar relativistic correction [91] has been considered in the study of CX_2 (X = F, Cl, and Br). To note, FCI energies at a given basis set can be approached using a scheme of correlation energy extrapolation by intrinsic scaling (CEEIS) [92–95], which has also been used in combination with additive corrections for core/valence correlation and scalar relativistic effects [78–82].

Despite the common use of CBS extrapolation by multireference methods, no systematic study of multireference composite approaches for excited states and potential energy curves have been performed until the introduction of MR-ccCA. Except for MR-ccCA and the multireference modifications of G2 and G3 [65], multireference methods have been used either solely based on icMRCI+Q or its computational equivalents such as MR-ACPF [96] and MR-AQCC [97], or solely based on multireference second-order perturbation theory (MRPT2) such as CASPT2. The aim of blending CASPT2 and icMRCI+Q (or its equivalents) in one composite approach was to retain the accuracy of icMRCI+Q and, at the same time, utilize the efficiency of CASPT2.

Unlike the hierarchy of single reference CC theory that progressively considers excitations higher than doubles (e.g. up to CCSDTQ5 has been applied for practical calculations), a generic MRCI algorithm taking account of excitation levels higher than doubles is computationally impractical. However, the inclusion of a multi-configuration reference space in multireference methods greatly improves the amount of higher order correlation being recovered as compared to single reference counterparts at the same level [98]. Consequently, the inclusion of triple and quadruple excitations in an MRCI frame might be sufficient for chemical accuracy or better. Conventionally, the "+Q" appending to MRCI indicates a posteriori Davidson's correction [99], which is a rough, albeit effective, estimation of correlation energy by quadruple excitations. MR-ACPF and MR-AQCC are regarded as improvements over MRCISD by considering individual disconnected triple and quadruple excitations effectively. Among the few multireference methods including connected triples and quadruples, nR-MRCISD(TQ) [100] is capable of considering several states simultaneously. An earlier benchmark study [100] demonstrated that nR-MRCISD(TQ) performs better than MR-ACPF and MR-AQCC. A highly accurate MRCI-based composite approach includes nR-MRCISD(TQ) as high-level correlation correction was proposed to obtain accurate potential energy curves for both ground and excited electronic states [63], which, in turn, can be used to benchmark the computationally more efficient composite approach MR-ccCA.

In the following sections, this report shows the formulation of MR-ccCA and its relationship to ccCA. The performance of several variants of MR-ccCA is compared and MR-ccCA is assessed by comparing to a MRCI-based composite approach or experimental data. Then, several applications of MR-ccCA are described. Finally, the report is concluded with the future developments and applications of multireference composite approaches.

2. FORMULATION OF MR-ccCA

Benchmark studies of composite approaches against reliable experimental data have predominantly been performed for single reference composite approaches [4]. In contrast, the effectiveness of CBS extrapolation schemes (see a recent review [98]) and additive corrections by multireference theories has been insufficiently examined despite the numerous recent studies using multireference methods. Assessing the performance of all possible combinations of multireference methods for each additive correction can be a cumbersome task. As a result, multireference analogs to the well-established ccCA were formulated initially by replacing the single reference theory in each composite step by a corresponding multireference theory [61]. Specifically, B3LYP and MP2 were replaced by CASPT2, and CCSD(T) was replaced by icMRCI+Q. However, some of the multireference theories have multiple competitive choices, e.g. the MR-ACPF or MR-AQCC replacement of icMRCI+Q, which are discussed in detail below. The representative theories employed within MR–ccCA are listed in Table 2.1 along with their single reference counterparts.

For thermochemical properties of molecules of three or more atoms, the geometry optimization and vibrational frequency calculations are performed by CASPT2 with the cc-pVTZ basis set, while the spectroscopic constants of diatomics are derived from the total MR–ccCA energies.

Similar to ccCA, the total MR–ccCA energy is based on a reference energy E_0(MR–ccCA) to which several corrections by multireference theory are added. The additive corrections include the higher-level dynamic correlation correction ΔE(CC), the outer core/valence correction ΔE(CV), the scalar relativistic correction ΔE(DK), and the spin–orbit coupling correction ΔE(SO).

$$E(\text{MR} - \text{ccCA}) = E_0(\text{MR} - \text{ccCA}) + \Delta E(\text{CC}) + \Delta E(\text{CV})$$
$$+ \Delta E(\text{DK}) + \Delta E(\text{SO}) \tag{2.1}$$

As pointed out by a recent review [98], the CBS extrapolation scheme for multireference methods remains ambiguous due in part to the difficulty in separating the correlation energy from the total energy. Again we borrowed

Table 2.1 The ccCA and MR-ccCA algorithms

	ccCA	MR-ccCA
Geometry optimization	B3LYP/cc-pVTZ	CASPT2/cc-pVTZ
ZVPE	B3LYP/cc-pVTZ★	CASPT2/cc-pVTZ
HF/CBS	HF/aug-cc-pVTZ	CASSCF/aug-cc-pVTZ
or	HF/aug-cc-pVQZ	CASSCF/aug-cc-pVQZ
CASSCF/CBS	$E(x) = E(\text{HF/CBS}) + A\exp(-1.63x)$	$E(x) = E(\text{CASSCF/CBS}) + a\exp(-1.63x)$
MP2/CBS	MP2/aug-cc-pVDZ	CASPT2/aug-cc-pVDZ
or	MP2/aug-cc-pVTZ	CASPT2/aug-cc-pVTZ
CASPT2/CBS	MP2/aug-cc-pVQZ	CASPT2/aug-cc-pVQZ
	"P": Eqn (2.3)	"P": Eqn (2.3)[†]
	"S3": Eqn (2.4)	"S3": Eqn (2.4)
	"S4": Eqn (2.5)	"S4": Eqn (2.5)
Correlation corrections	CCSD(T)/cc-pVTZ—MP2/cc-pVTZ	icMRCI+Q^c/cc-pVTZ—CASPT2/cc-pVTZ
Core/valence corrections	MP2(FC1)/aug-cc-pCVTZ— MP2/aug-cc-pVTZ	CASPT2(FC1)/aug-cc-pCVTZ— CASPT2/aug-cc-pCVTZ
Scalar relativistic correction	MP2/cc-pVTZ-DK - MP2/cc-pVTZ	CASPT2/cc-pVTZ-DK - CASPT2/cc-pVTZ
Spin–orbit corrections	Experimental values for atoms when applicable	Experimental values for atoms when applicable

★ Experimental values for atoms and harmonic frequencies scaled by 0.989 for molecules.
[†] The total CASPT2 energies were used by Eqn (2.3).
MR-ACPF and MR-AQCC may be used instead of icMRCI+Q.

the strategy from ccCA to calculate the reference energy E_0(MR-ccCA). CASSCF and subsequent CASPT2 calculations are performed with the aug-cc-pVxZ ($x = 2$ or D, 3 or T, and 4 or Q) [85,101] series of basis sets. Then the CASSCF energy is extrapolated similar to the HF energy by using an exponential equation [102–104] or an effective two-point extrapolation of energies by triple-ζ (TZ) and quadruple-ζ (QZ) basis sets [105].

$$E(x) = E(CBS) + a\exp(-1.63x) \tag{2.2}$$

where a is a parameter to be determined by the energy points.

Similar to the MP2 correlation energy extrapolations, three different schemes are considered for the dynamical correlation energies of CASPT2. The CASPT2 energies are obtained by a three-point extrapolation using the mixed Gaussian formula [106]

$$E(x) = E(CBS) + A\exp\left[-(x-1)\right] + B\exp\left[-(x-1)^2\right] \tag{2.3}$$

or the two-point extrapolation using the inverse cubic power of x [107,108]

$$E(x) = E(CBS) + Ax^{-3} \tag{2.4}$$

or the inverse quartic power of $(x + 1/2)$

$$E(x) = E(CBS) + A\left(x + \frac{1}{2}\right)^{-4}. \tag{2.5}$$

The total CASPT2 energy including the reference CASSCF energy has been extrapolated using Eqn (2.3) and the resulting method is termed MR-ccCA-P. Alternatively, "PII," "S3," and "S4" appending to MR-ccCA specify the extrapolation Eqns (2.3)–(2.5), respectively, for CASPT2 dynamical correlation energies and Eqn (2.2) for the reference CASSCF energy. PI is the same as PII except that the exponential (-1.63 for PII) in Eqn (2.2) is relaxed. And "PS3" specifies that an averaged value of the P and S3 extrapolations is considered [15].

The dynamic correlation beyond CASPT2 is calculated by icMRCI with the a posteriori Davidson's correction ($+Q$) [99,109], MR-ACPF, or MR-AQCC. The latter two include a priori size consistency corrections.

$$\Delta E(CC) = icMRCI + Q/cc - pVTZ - CASPT2/cc - pVTZ \tag{2.6}$$

$$\Delta E(CC) = MR - ACPF/cc - pVTZ - CASPT2/cc - pVTZ \tag{2.7}$$

$$\Delta E(CC) = MR - AQCC/cc - pVTZ - CASPT2/cc - pVTZ \tag{2.8}$$

ACPF or AQCC are appended to MR–ccCA when Eqn (2.7) or (2.8) are used within MR–ccCA to differentiate from the default MR–ccCA using Eqn (2.6). To note, it is ideal to use rigorously size consistent MR–CCSD [110–113] or MR–CCSD(T) for high-level dynamic correlation correction. However, these methods are either still in the development stage or not readily available in current popular quantum programs.

The core/valence corrections are considered by CASPT2 using the aug-cc-pCVTZ basis set [88]. "FC1" indicates that the outer core electrons are correlated, i.e. 1s electrons for first row elements Li–Ne, 2s2p electrons for second row Na–Ar, 3s3p electrons for K–Kr, and 3s3p electrons for $3d$ transition metals Sc–Zn are correlated in addition to the valence electrons.

$$\Delta E(CV) = CASPT2(FC1)/aug - cc - pCVTZ - CASPT2/aug \\ - cc - pCVTZ \tag{2.9}$$

The scalar relativistic correction is obtained by employing the spin free one-electron Douglas–Kroll–Hess Hamiltonian operator in combination with the recontracted DK basis set [90].

$$\Delta E(DK) = CASPT2/cc - pVTZ - DK - CASPT2/cc - pVTZ \tag{2.10}$$

The different modifications of CASPT2 (the default, g1, g2, g3, and g4 [114]) as implemented in MOLPRO [115] were compared. There are also other formulations of MRPT2 than CASPT2 (e.g. RASPT2 [116], NEVPT2 [117–119], MRMP2 [120,121], GVVPT2 [122,123], to name only a few) and different implementations of multireference methods than those in MOLPRO. Although a comparison and discussion of these alternative methods or implementations would be useful, it is beyond the scope of this report.

3. FORMULATION OF MRCI-BASED COMPOSITE APPROACH

Due to the scarcity of accurate potential energies for both ground and excited states through the entire geometries from equilibrium to dissociated atoms, an MRCI-based composite approach was suggested to construct highly accurate potential energy curves. The total composite energy includes a reference energy by icMRCI+Q, core/valence corrections, scalar relativistic corrections, and high-level correlation correction by nR-MRCISD(TQ).

$$E = E_0 + \Delta E(\text{CV}) + \Delta E(\text{DK}) + \Delta E(\text{TQ}) \qquad (2.11)$$

where E_0 is the extrapolated CBS limit of the icMRCI+Q/aug-cc-pVxZ ($x = 5, 6$) energies and

$$\Delta E(\text{CV}) = \text{icMRCI} + \text{Q(FC1)}/\text{aug} - \text{cc} - \text{pCV}x\text{Z} - \text{icMRCI}$$

$$+ \text{Q}/\text{aug} - \text{cc} - \text{pCV}x\text{Z} \quad (x = 4 \text{ or } \text{Q, and } 5)$$

$$(2.12)$$

$$\Delta E(\text{DK}) = \text{icMRCI} + \text{Q(FC1)}/\text{aug} - \text{cc} - \text{pCVTZ} - \text{DK} - \text{icMRCI}$$

$$+ \text{Q(FC1)}/\text{aug} - \text{cc} - \text{pCVTZ}$$

$$(2.13)$$

$$\Delta E(\text{TQ}) = \text{nR} - \text{MRCISD(TQ)}/\text{aug} - \text{cc} - \text{pVTZ} - \text{icMRCI}$$

$$+ \text{Q}/\text{aug} - \text{cc} - \text{pVTZ}$$

$$(2.14)$$

Equations (2.2) and (2.3) are used for reference (CASSCF) and dynamical correlation energies, respectively, in all CBS extrapolations within the MRCI-based composite approach.

4. PREDICTABILITY USING MR-ccCA

MR-ccCA can be used for chemical situations such as bond-formation and bond-breaking processes, transition states, and excited electronic states where single reference methods may fail to give a qualitatively correct description. Below, an overview of examples demonstrating the utility of MR-ccCA has been provided (These examples are detailed further in refs. 61–64). An important aspect of these early studies has been the evaluation of MR-ccCA and the various possible forms of MR-ccCA. To gauge this, the performance of MR-ccCA has been compared to either experimental data or theoretical predictions by a highly accurate composite approach based on MRCI methods.

4.1. Potential Energy Curves of C_2

The potential energy curves of the $X^1\Sigma_g^+$, $B^1\Delta_g$, and $B'^1\Sigma_g^+$ states of C_2 are challenging for quantum mechanical theory due to the crossing between the $B^1\Delta_g$ and $B'^1\Sigma_g^+$ states at the atomic distance of ~ 1.1 Å, a crossing between $X^1\Sigma_g^+$ and $B^1\Delta_g$, and an avoided crossing between $X^1\Sigma_g^+$ and $B'^1\Sigma_g^+$ at 1.7 Å [58]. To assess the quality of MR–ccCA for geometries far away from equilibrium, the MRCI-based composite approach, nR–MRCISD(TQ), was used to construct the potential energy curves. nR–MRCISD(TQ) predicts equilibrium bond length r_e, harmonic frequency ω_e, and term energy (or the difference between the energy minima of an excited state and the ground state) T_e with mean absolute deviations (MAD) of 0.0002 Å, $0.7\,\mathrm{cm}^{-1}$, and $37\,\mathrm{cm}^{-1}$ ($=0.11\,\mathrm{kcal\,mol}^{-1}$), respectively, from experiments (To note, a chemical accuracy of $\pm 1.0\,\mathrm{kcal\,mol}^{-1}$ is expected for T_e). This approach meets the stringent requirement of spectroscopic accuracy of ± 0.0005 Å for structure and $\pm 1\,\mathrm{cm}^{-1}$ for vibrational frequency [4]. The predicted dissociation energies (D_e) are -0.4, 0.7, and $0.5\,\mathrm{kcal\,mol}^{-1}$ away from the experimental values (145.8, 110.7, and $101.2\,\mathrm{kcal\,mol}^{-1}$) for the $X^1\Sigma_g^+$, $B^1\Delta_g$, and $B'^1\Sigma_g^+$ states of C_2, respectively, and the deviations from experiments are within the chemical accuracy of $\pm 1.0\,\mathrm{kcal\,mol}^{-1}$. The high accuracy of the MRCI-based approach assures that the obtained potential energies can be used as a benchmark for the more efficient MR–ccCA for all geometries spanning from equilibrium to dissociation.

The definition of the zeroth-order Hamiltonian in CASPT2 is not unique and there are four modifications (g1–g4) [114] available in MOL-PRO [115]. Essentially the same MR–ccCA results were obtained with each of the four modifications of CASPT2. Both g3 and g4 are legitimate choices in the MR–ccCA formulation. The deviations of MR–ccCA-P energies (relative to equilibrium energies) from the corresponding relative energies by the MRCI-based composite method are less than $2.0\,\mathrm{kcal\,mol}^{-1}$. This result was surprising as the additive corrections within MR–ccCA work synergistically even at the geometries where the complicated crossing and avoided crossing of multiple electronic states occur. This supports that MR–ccCA can be a viable approach for excited states and accurate potential energy curves that depict the bond-formation and bond-breaking processes. MR–ccCA-P predicts the r_e, ω_e, T_e, and D_e with MADs from experiments of 0.0007 Å, $3.5\,\mathrm{cm}^{-1}$, $200\,\mathrm{cm}^{-1}$ ($=0.57\,\mathrm{kcal\,mol}^{-1}$), and $0.7\,\mathrm{kcal\,mol}^{-1}$, respectively, for the three states of C_2. Similar accuracy was achieved for the ground state of N_2 for r_e, ω_e, T_e, but a large deviation of $-1.8\,\mathrm{kcal\,mol}^{-1}$ ($-1.4\,\mathrm{kcal\,mol}^{-1}$ in an earlier study [61]) from experiments was observed with MR–ccCA-P.

4.2. Spectroscopic Constants of C_2, N_2, and O_2

MR–ccCA has been used to calculate the spectroscopic constants of the ground and valence excited states of diatomics C_2, N_2, and O_2 for a systematic assessment by comparing to reliable experimental data. 15 singlet and 14 triplet excited electronic states were considered for the statistical analysis after excluding a few valence excited states with unreliable experimental data or possible strong interactions with Rydberg states which require active orbitals beyond the valence shell. The overall MR–ccCA-P MADs are 0.0006 Å, 7.0 cm^{-1}, and 143 cm^{-1} ($=0.41$ kcal mol^{-1}) for r_e, ω_e, and T_e respectively, showing a better accuracy than MR–AQCC [124] with the CBS extrapolation of the cc-pVTZ and cc-pVQZ basis sets and an estimated core/valence correction. Similar results, i.e. 0.0006–0.0008 Å, 6.1–7.0 cm^{-1}, and 143–154 cm^{-1} for the MADs of r_e, ω_e, and T_e respectively [63], were obtained for all extrapolation schemes P, S3, S4, and PS3 as outlined in Section 2.2. On average, MR–ccCA results are within the chemical accuracy for spectroscopic properties, i.e. ± 0.005 Å and ± 15 cm^{-1} for geometries and frequencies, respectively, as recently defined by Peterson, Feller, and Dixon [4]. The MR–ccCA-P MAD for D_e is 1.06 kcal mol^{-1}, close to the chemical accuracy of ± 1.0 kcal mol^{-1} for atomization energies.

Replacing Davidson's correction to icMRCI by Pople's correction [125,126] within MR–ccCA increases the MADs of T_e and D_e from 143 cm^{-1} and 1.06 kcal mol^{-1} to 190 cm^{-1} and 2.00 kcal mol^{-1}, respectively, but gives similar MADs for r_e (0.0006 versus 0.0006 Å) and ω_e (7.0 versus 6.5 cm^{-1}) [63], which are less prone to size consistency error than T_e and D_e. Even though MR–ccCA-P with Pople's correction yielded much larger MADs for the states of C_2 and O_2 than MR–ccCA-P with Davidson's correction, the former gave a MAD of 0.67 kcal mol^{-1} relative to experiments for the states of N_2, much less than the MAD of 1.49 kcal mol^{-1} by the latter. The CASPT2 calculations with correlated core electrons may affect the MR–ccCA atomic energies substantially. For example, the MR–ccCA-P MAD of D_e is 1.98 kcal mol^{-1} for the states of N_2 when *single* excitations are *internally contracted* in the CASPT2(full)/aug-cc-pCVTZ calculations for atomic energies. By relaxing the single excitations in the CASPT2(full)/aug-cc-pCVTZ step, the MR–ccCA-P calculations predicted a relative atomic energy for the ^{2}D state of N atom with a much lower deviation of only 0.1 kcal mol^{-1} from the experimental value. However, relaxing single excitations in CASPT2(full)/aug-cc-pCVTZ increases the errors relative to experiments for the atomic energies of C atom. A further investigation is warranted to resolve the unusual discrepancy by relaxing/ contracting the single excitations in the core electrons correlated CASPT2 calculations for atoms.

4.3. TAEs and Enthalpies of Formation

The successes of MR-ccCA in predicting potential energies of both ground and excited electronic states with chemical accuracy allowed us to extend MR-ccCA to the calculations of thermodynamic properties [i.e. TAEs or enthalpies of formation (ΔH_f)] for species [62,64] that may possess multireference character so strong that single reference composite methods would fail to give even qualitatively correct results. The calculations of TAEs (or ΔH_f) were performed *via* atomistic approach, which is subject to the size consistency error to a much greater extent than properties derived from potential energies calculated for the whole molecule. As shown in previous paragraphs, the size consistency corrections are particularly important for the quality of MR-ccCA predictions of the dissociation energy D_e, which is pertinent to TAE. Among the various size consistency corrections, the a priori size consistency corrected methods MR-ACPF and MR-AQCC have been widely used, and they were employed in place of icMRCI+Q within MR-ccCA. Our initial test calculations showed that the difference between MR-ACPF and MR-AQCC are trivial but both are slightly better than icMRCI+Q for ΔH_fs. Consequently, both MR-ACPF and MR-AQCC have been used within MR-ccCA for the study of TAEs (or ΔH_fs), but the specific theory is not listed for clarity.

The MR-ccCA study of TAEs (or ΔH_fs) for a number of second row and third row species with diverse electronic structure is surveyed (Table 2.2). Except for Si_2H and Si_2H_3, whose TAEs or ΔH_fs are experimentally unknown, all species are listed with experimental data including some with large uncertainties. To note, many of the molecules have significant multireference character as indicated by the diagnostic $T_1 > 0.02$, $D_1 > 0.05$ [127], and $C_0^2 < 0.90$. These species may require a multi-configurational reference wavefunction for quantitatively correct description and sophisticated single reference composite methods such as W4 and HEAT or CC theory with high-levels of excitations may be needed for reliable theoretical TAEs. Currently, it is difficult to statistically assess the quality of MR-ccCA for the predictions of TAEs (or ΔH_f) for multireference systems due to a scarcity of reliable experimental or theoretical reference data. Nonetheless, MR-ccCA predicted that ΔH_fs are within the chemical accuracy of

Table 2.2 Enthalpies of formation (ΔH_f) and total atomization energies (TAE) for the ground state of first and second row species (kcal mol^{-1}) [62,64]

Molecule	MR-ccCA		Experimental*	
	ΔH_f	TAE	ΔH_f	TAE
Homologous dimers and trimers				
C_2	199.1	142.4	200.1	
N_2		228.4†		227.0†
O_2	0.4		0.0	
Si_2	145.6	73.1	141.0	76.2 ± 1.7
S_2	31.6		30.7 ± 0.1	
C_3	196.6	314.7	196.0	314.4 ± 3.1
O_3	35.4		34.1 ± 0.4	
Si_3	158.0	169.7	156.1 ± 3.8	168.5 ± 3.8
S_3	36.3		33 ± 3	
Hydrides				
CH	142.5	79.9	142.8	79.9
CH_2	94.4		93.7	
C_2H	137.6	255.0	135.8 ± 1.4	257.5
C_2H_2	56.2		54.3 ± 0.2	
C_2H_3	72.8	421.1	72.1	422.2
C_2H_4	12.6		12.5	
SiH	90.5	70.4	90.0	68.7 ± 0.7
SiH_2	64.9		65.2	
Si_2H	122.5	147.1		
Si_2H_2	96.6	222.8	<99.7	
Si_2H_3	99.9	271.3		
Si_2H_4	66.8		65.7 ± 0.9	
Others				
CF_2	−44.5		$−43.5 \pm 1.5$	
CHCl	77.4		80.4 ± 2.8	
CCl_2	57.5		55.0 ± 2.0	
CHF	35.6		34.2 ± 3.0	
CH_2O	−25.6		−26.0	
SO	3.1		1.2	
SiB	172.3	74.3	166.8 ± 3.3	74.6 ± 2.8
Si_2B	173.6	181.8	164.4 ± 4.7	183.3 ± 4.3
SiC_2	157.3	293.1		301.0 ± 7
Si_2C	139.5	249.3		256.1 ± 6
Si_2N	99.3	231.5	85.0 ± 3.5	241.1 ± 3.0

*See Oyedepo [62] and Wilson [64] for the sources of experimental data.
†Dissociation energy D_e.

$\pm 1.0\,\mathrm{kcal\,mol}^{-1}$ for most species with reliable experimental data (see Table 2.2). MR–ccCA predictions can be compared to the experimental data with large uncertainty and determine if it is necessary to revisit the experimental data.

Besides the species listed in Table 2.2, MR–ccCA has been used to predict TAEs and ΔH_fs for other species whose energetics may be experimentally unknown. For examples, the diatomic (SiX, X = B, C, N, Al, and P) and triatomic (SiX$_2$ and Si$_2$X, X = B, C, N, Al, and P) compounds of silicon were investigated and some of them have never been experimentally reported or theoretically studied prior to our study. For molecules of more than two atoms, it is impractical to scan the potential energy surface by MR–ccCA. CASPT2/cc-pVTZ with a balanced treatment of nondynamical correlation and significant amount of dynamical correlation has been used to optimize the geometries and calculate the vibrational frequencies. The CASPT2/cc-pVTZ geometries along with the MR–ccCA TAEs were reported for the new silicon species. The MR–ccCA calculations of ΔH_fs have also been performed for the excited states. However, the experimental ΔH_f is seldom reported for excited states. Instead, the adiabatic transition energies by MR–ccCA have been compared to available experimental data and/or other theoretical calculations as discussed in next section.

4.4. Adiabatic Transition Energies

Another important application of MR–ccCA is the calculation of adiabatic transition energies of excited states. In addition to the valence excited states of C$_2$, N$_2$, and O$_2$, the lowest spin-forbidden transitions have been considered for atoms and a few first row and second row species with diverse electronic structure. We compared the use of icMRCI+Q, MR-AQCC, and MR-ACPF as high-level correlation theory within MR–ccCA for the singlet–triplet energy separations for atoms C, O, Si, and S. While MR-AQCC and MR-ACPF are similar in performance, both are better than icMRCI+Q in accuracy. MR–ccCA-P with MR-AQCC predicted the energy separations with deviations of 0.3, 0.5, 0.3, and 0.7 kcal mol^{-1} from experiments. For the selected 17 molecules (Table 2.3), the deviations of MR–ccCA energy separations are within $\pm 1.0\,\mathrm{kcal\,mol}^{-1}$ of experimental data except for SiH$_2$ ($-1.4\,\mathrm{kcal\,mol}^{-1}$), NF ($-1.4\,\mathrm{kcal\,mol}^{-1}$), NH$_2^+$ ($-2.0\,\mathrm{kcal\,mol}^{-1}$), PH$_2^+$ ($-1.5\,\mathrm{kcal\,mol}^{-1}$), and the overall MAD is $0.65\,\mathrm{kcal\,mol}^{-1}$.

Table 2.3 Transition energies for the lowest spin-forbidden excited states (kcal mol^{-1}) [62,64]

Molecule	MR-ccCA	Experimental*	Deviation
CH_2	8.2	9.0	−0.8
SiH_2	−22.4	−21.0 ± 0.7	−1.4
O_2	22.6	22.6	0.0
NH	36.6	35.9	0.7
OH^+	50.4	50.5	−0.1
NF	32.9	34.3	−1.4
SO	17.1	16.8	0.3
S_2	12.6	12.6	0.0
O_3	29.0	28.6	0.4
NH_2^+	28.1	30.1 ± 0.2	−2.0
PH_2^+	−18.8	−17.3	−1.5
CF_2	−56.3	−56.7	0.4
CHCl	−5.7	−6.2	0.5
CHF	−15.1	−14.9 ± 0.4	−0.2
CH	17.6	17.1 ± 0.2	0.5
C_2	1.3	2.1	−0.8
Si_2	9.9	10.0	−0.1

*See Oyedepo [62] and Wilson [64] for the sources of experimental data.

4.5. Reaction Barrier Heights

Competitive configurations may exist in the transition state of reactions, and thus MR–ccCA is deemed an appropriate method for locating the transition structure (via CASPT2/cc-pVTZ) and accurately computing reaction barrier heights. Single reference methods, except those including CC methods with excitations higher than triples, fail to predict the correct barrier height of many reactions quantitatively, e.g. ccCA gives an unphysical hump on the potential energies of C_2H_4 with respect to the torsional angle [62]. However, MR-ccCA predicted the barrier height as 64.3 kcal mol^{-1} [62], confirming an experimental estimation of 65 kcal mol^{-1} [128] for the torsional rotation of C_2H_4. For the heavier valence isoelectronic Si_2H_4, the MR-ccCA barrier height of the torsional rotation is 25.5 kcal mol^{-1}, in excellent corroboration with the experimental value of 25.8 ± 4.8 kcal mol^{-1} [129].

4.6. Future of MR-ccCA

Given the success of MR–ccCA as an extension of ccCA for chemical situations that require a multireference description, the application of MR-ccCA can be extended to other chemical systems of salient multireference character. However, MR-ccCA is subject to the limitations of the constituent methods, e.g. the determination of active space. In our early calculations, a full valence active space including all valence electrons and valence orbitals of constituent atoms has been employed. However, such an active space will cause an unfavorable scaling of computational cost as the number of atoms increases. It is possible to reduce the size of active space by using a small complete active space and/or a restricted active space. High-level correlation methods icMRCI, MR-AQCC, and MR-ACPF may not recover sufficient correlation energy, particularly dynamical correlation energy, as they do not include excitations or connected excitations higher than doubles. It is highly desirable to include a multireference equivalent to CCSD(T), e.g. MR-CCSD(T), in MR-ccCA when it becomes readily available.

5. CONCLUSIONS AND REMARKS

The single reference ccCA and its variants have found vast successful applications in predicting accurate energetics and other properties for both main group and transition metal species. The ccCA module has been implemented in popular chemical programs such as GAUSSIAN [130] and NWChem [131], and for other programs such as MOLPRO and ccCA scripts are available from the authors upon request. The multireference analog of ccCA, or MR-ccCA, has been formulated by utilizing CASPT2 and icMRCI+Q (or MR-AQCC or MR-ACPF) analogous to MP2 and CCSD(T) within single reference ccCA. The CBS extrapolation, core/valence corrections, and scalar relativistic effects are evaluated by CASPT2, and savings in computational cost are achieved by using the effective CASPT2 instead of the more costly MRCI (or MR-AQCC or MR-ACPF). The correlation correction is done by icMRCI+Q (or MR-AQCC or MR-ACPF), but MR-ccCA is more efficient as the cc-pVTZ basis set of a moderate size is utilized. The Peterson extrapolation scheme is recommended for CASPT2/CBS energies within MR-ccCA although other extrapolation schemes that we have considered such as S3 and S4 also give similar results. Each individual step in MR-ccCA can be easily realized in commonly programs such as MOLPRO [115] as long as the constituent methods are available.

MR-ccCA can predict reliable and accurate potential energies of complicated electronic states at geometries spanning from equilibrium to dissociated limit, which is substantiated by the close agreement of MR-ccCA energies (relative to the equilibrium energies) with those by a highly accurate MRCI-based approach for electronic states of C_2. MR-ccCA has achieved accurate predictions for spectroscopic properties [4], i.e. ± 0.005 Å for geometries and ± 15 cm^{-1} for vibrational frequencies, in the study of valence excited states of C_2, N_2, and O_2. For the term energy T_e and dissociation energy D_e, MR-ccCA MADs from experiments are 1.06 and 0.41 kcal mol^{-1}, respectively. MR-ccCA has also attained good to excellent accuracy for other energetic properties such as TAEs, enthalpies of formation, transition energies, and reaction barrier heights. In summary, the CASPT2-based MR-ccCA can be used as an effective multireference method for spectroscopic, thermodynamic, and kinetic properties when chemical accuracy is sought.

ACKNOWLEDGMENTS

The authors gratefully acknowledge the support from the National Science Foundation (grant no. CHE-0809762). Grateful acknowledgments also go to the United States Department of Education for the support of the Center for Advanced Scientific Computing and Modeling (CASCaM).

REFERENCES

[1] Raghavachari, K.; Trucks, G. W.; Pople, J. A.; Head-Gordon, M. *Chem. Phys. Lett.* **1989**, *157*, 479.

[2] Helgaker, T.; Klopper, W.; Tew, D. P. *Mol. Phys.* **2008**, *106*, 2107.

[3] Gordon, M. S.; Mullin, J. M.; Pruitt, S. R.; Roskop, L. B.; Slipchenko, L. V.; Boatz, J. A. *J. Phys. Chem. B* **2009**, *113*, 9646.

[4] Peterson, K.; Feller, D.; Dixon, D. *Theor. Chem. Acc.* **2012**, *131*, 1079.

[5] John, A. P.; Martin, H.-G.; Douglas, J. F.; Krishnan, R.; Larry, A. C. *J. Chem. Phys.* **1989**, *90*, 5622.

[6] Curtiss, L. A.; Jones, C.; Trucks, G. W.; Raghavachari, K.; Pople, J. A. *J. Chem. Phys.* **1990**, *93*, 2537.

[7] Curtiss, L. A.; Raghavachari, K.; Trucks, G. W.; Pople, J. A. *J. Chem. Phys.* **1991**, *94*, 7221.

[8] Curtiss, L. A.; Raghavachari, K.; Redfern, P. C.; Rassolov, V.; Pople, J. A. *J. Chem. Phys.* **1998**, *109*, 7764.

[9] Curtiss, L. A.; Redfern, P. C.; Raghavachari, K.; Rassolov, V.; Pople, J. A. *J. Chem. Phys.* **1999**, *110*, 4703.

[10] Baboul, A.; Curtiss, L.; Redfern, P.; Raghavachari, K. *J. Chem. Phys.* **1999**, *110*, 7650.

[11] Curtiss, L. A.; Redfern, P. C.; Raghavachari, K. *J. Chem. Phys.* **2007**, *126*, 084108.

[12] Curtiss, L. A.; Redfern, P. C.; Raghavachari, K. *J. Chem. Phys.* **2007**, *127*, 124105.

[13] DeYonker, N. J.; Cundari, T. R.; Wilson, A. K. *J. Chem. Phys.* **2006**, *124*, 114104.

[14] DeYonker, N. J.; Grimes, T.; Yockel, S.; Dinescu, A.; Mintz, B.; Cundari, T. R.; Wilson, A. K. *J. Chem. Phys.* **2006**, *125*, 104111.

[15] DeYonker, N. J.; Wilson, B. R.; Pierpont, A. W.; Cundari, T. R.; Wilson, A. K. *Mol. Phys.* **2009**, *107*, 1107.

[16] Grimes, T. V.; Wilson, A. K.; DeYonker, N. J.; Cundari, T. R. *J. Chem. Phys.* **2007**, *127*, 154117.

[17] Boese, A. D.; Oren, M.; Atasoylu, O.; Martin, J. M. L.; Kallay, M.; Gauss, J. *J. Chem. Phys.* **2004**, *120*, 4129.

[18] Karton, A.; Rabinovich, E.; Martin, J. M. L.; Ruscic, B. *J. Chem. Phys.* **2006**, *125*, 144108.

[19] Martin, J. M. L.; de Oliveira, G. *J. Chem. Phys.* **1999**, *111*, 1843.

[20] Karton, A.; Daon, S.; Martin, J. M. L. *Chem. Phys. Lett.* **2011**, *510*, 165.

[21] Harding, M. E.; Vazquez, J.; Ruscic, B.; Wilson, A. K.; Gauss, J.; Stanton, J. F. *J. Chem. Phys.* **2008**, *128*, 114111.

[22] Bomble, Y. J.; Vazquez, J.; Kallay, M.; Michauk, C.; Szalay, P. G.; Csaszar, A. G.; Gauss, J.; Stanton, J. F. *J. Chem. Phys.* **2006**, *125*, 064108.

[23] Tajti, A.; Szalay, P. G.; Csaszar, A. G.; Kallay, M.; Gauss, J.; Valeev, E. F.; Flowers, B. A.; Vazquez, J.; Stanton, J. F. *J. Chem. Phys.* **2004**, *121*, 11599.

[24] East, A. L. L.; Johnson, C. S.; Allen, W. D. *J. Chem. Phys.* **1993**, *98*, 1299.

[25] East, A. L. L.; Allen, W. D. *J. Chem. Phys.* **1993**, *99*, 4638.

[26] Wood, G. P. F.; Radom, L.; Petersson, G. A.; Barnes, E. C.; Frisch, M. J.; Montgomery, J. A. *J. Chem. Phys.* **2006**, *125*, 094106.

[27] Nyden, M. R.; Petersson, G. A. *J. Chem. Phys.* **1981**, *75*, 1843.

[28] Petersson, G. A.; Al-Laham, M. A. *J. Chem. Phys.* **1991**, *94*, 6081.

[29] Petersson, G. A.; Tensfeldt, T. G.; Montgomery, J. J. A. *J. Chem. Phys.* **1991**, *94*, 6091.

[30] Montgomery, J. J. A.; Ochterski, J. W.; Petersson, G. A. *J. Chem. Phys.* **1994**, *101*, 5900.

[31] Ochterski, J. W.; Petersson, G. A.; Montgomery, J. J. A. *J. Chem. Phys.* **1996**, *104*, 2598.

[32] Montgomery, J. J. A.; Frisch, M. J.; Ochterski, J. W.; Petersson, G. A. *J. Chem. Phys.* **1999**, *110*, 2822.

[33] Montgomery, J. J. A.; Frisch, M. J.; Ochterski, J. W.; Petersson, G. A. *J. Chem. Phys.* **2000**, *112*, 6532.

[34] Feller, D.; Peterson, K. A.; Dixon, D. A. *J. Chem. Phys.* **2008**, *129*, 204105.

[35] Feller, D.; Peterson, K. A. *J. Chem. Phys.* **2007**, *126*, 114105.

[36] Feller, D. *J. Chem. Phys.* **1993**, *98*, 7059.

[37] Feller, D.; Dixon, D. A. *J. Phys. Chem. A* **2000**, *104*, 3048.

[38] Dixon, D. A.; Feller, D.; Peterson, K. A. *J. Chem. Phys.* **2001**, *115*, 2576.

[39] Dunning, T. H., Jr.; Peterson, K. A. *J. Chem. Phys.* **2000**, *113*, 7799.

[40] Feller, D.; Peterson, K. A.; Crawford, T. D. *J. Chem. Phys.* **2006**, *124*, 054107.

[41] Fast, P. L.; Corchado, J. C.; Sanchez, M.; Truhlar, D. G. *J. Phys. Chem. A* **1999**, *103*, 5129.

[42] Fast, P. L.; Truhlar, D. G. *J. Phys. Chem. A* **2000**, *104*, 6111.

[43] Tasi, G.; Izsak, R.; Matisz, G.; Csaszar, A. G.; Kallay, M.; Ruscic, B.; Stanton, J. F. *Chem. Phys. Chem.* **2006**, *7*, 1664.

[44] Ruscic, B.; Boggs, J. E.; Burcat, A.; Csaszar, A. G.; Demaison, J.; Janoschek, R.; Martin, J. M. L.; Morton, M. L.; Rossi, M. J.; Stanton, J. F.; Szalay, P. G.; Westmoreland, P. R.; Zabel, F.; Berces, T. *J. Phys. Chem. Ref. Data* **2005**, *34*, 573.

[45] Ruscic, B.; Michael, J. V.; Redfern, P. C.; Curtiss, L. A.; Raghavachari, K. *J. Phys. Chem. A* **1998**, *102*, 10889.

[46] Ruscic, B.; Pinzon, R. E.; Morton, M. L.; Srinivasan, N. K.; Su, M.-C.; Sutherland, J. W.; Michael, J. V. *J. Phys. Chem. A* **2006**, *110*, 6592.

[47] Feller, D. *J. Comput. Chem.* **1996**, *17*, 1571.

[48] Curtiss, L. A.; Redfern, P. C.; Raghavachari, K. *J. Chem. Phys.* **2005**, *123*, 124107.

[49] DeYonker, N. J.; Mintz, B.; Cundari, T. R.; Wilson, A. K. *J. Chem. Theory Comput.* **2008**, *4*, 328.

[50] DeYonker, N. J.; Ho, D. S.; Wilson, A. K.; Cundari, T. R. *J. Phys. Chem. A* **2007**, *111*, 10776.

[51] Ho, D. S.; DeYonker, N. J.; Wilson, A. K.; Cundari, T. R. *J. Phys. Chem. A* **2006**, *110*, 9767.

[52] Jiang, W.; DeYonker, N. J.; Determan, J. J.; Wilson, A. K. *J. Phys. Chem. A* **2011**, *116*, 870.

[53] DeYonker, N. J.; Williams, T. G.; Imel, A. E.; Cundari, T. R.; Wilson, A. K. *J. Chem. Phys.* **2009**, *131*, 024106.

[54] DeYonker, N. J.; Peterson, K. A.; Steyl, G.; Wilson, A. K.; Cundari, T. R. *J. Phys. Chem. A* **2007**, *111*, 11269.

[55] Laury, M. L.; DeYonker, N. J.; Jiang, W.; Wilson, A. K. *J. Chem. Phys.* **2011**, *135*, 214103.

[56] Schmidt, M. W.; Gordon, M. S. *Annu. Rev. Phys. Chem.* **1998**, *49*, 233.

[57] Musiał, M.; Bartlett, R. J. *J. Chem. Phys.* **2005**, *122*, 224102.

[58] Abrams, M. L.; Sherrill, C. D. *J. Chem. Phys.* **2004**, *121*, 9211.

[59] Piecuch, P.; Wloch, M. *J. Chem. Phys.* **2005**, *123*, 224105.

[60] Nedd, S. A.; DeYonker, N.J.; Wilson, A. K.; Piecuch, P.; Gordon, M. S. *J. Chem. Phys.* **2012**, *136*, 144109.

[61] Mintz, B.; Williams, T. G.; Howard, L.; Wilson, A. K. *J. Chem. Phys.* **2009**, *130*, 234104.

[62] Oyedepo, G. A.; Wilson, A. K. *J. Phys. Chem. A* **2010**, *114*, 8806.

[63] Jiang, W.; Wilson, A. K. *J. Chem. Phys.* **2011**, *134*, 034101.

[64] Oyedepo, G. A.; Peterson, C.; Wilson, A. K. *J. Chem. Phys.* **2011**, *135*, 094103.

[65] Sølling, T. I.; Smith, D. M.; Radom, L.; Freitag, M. A.; Gordon, M. S. *J. Chem. Phys.* **2001**, *115*, 8758.

[66] Martin, J. M. L.; Parthinban, S. In *Quantum Mechanical Prediction of Thermochemical Data*; Cioslowski, J., Szarecka, A., Eds.; Kluwar Academic Publishers: Dordrecht, The Netherlands, 2001; pp 31–65.

[67] Mintz, B.; Chan, B.; Sullivan, M. B.; Buesgen, T.; Scott, A. P.; Kass, S. R.; Radom, L.; Wilson, A. K. *J. Phys. Chem. A* **2009**, *113*, 9501.

[68] Shi, D.-H.; Xing, W.; Sun, J.-F.; Zhu, Z.-L.; Liu, Y.-F. *Int. J. Quantum Chem.* **2012**, *112*, 1323.

[69] Shi, D.; Li, W.; Sun, J.; Zhu, Z.; Liu, Y. *Comput. Theor. Chem.* **2012**, *978*, 126.

[70] Shi, D.; Li, W.; Sun, J.; Zhu, Z. *Spectrochim. Acta, Part A* **2012**, *87*, 96.

[71] Shi, D.-H.; Liu, H.; Sun, J.-F.; Zhu, Z.-L.; Liu, Y.-F. *J. Quant. Spectrosc. Radiat. Transfer* **2011**, *112*, 2567.

[72] Shi, D.; Zhang, X.; Sun, J.; Zhu, Z. *Mol. Phys.* **2011**, *109*, 1453.

[73] Peterson, K. A. *Mol. Phys.* **2010**, *108*, 393.

[74] Song, Y. Z.; Varandas, A. J. C. *J. Phys. Chem. A* **2011**, *115*, 5274.

[75] Varandas, A. J. C. *J. Chem. Phys.* **2009**, *131*, 124128.

[76] Varandas, A. J. C. *J. Chem. Phys.* **2008**, *129*, 234103.

[77] Varandas, A. J. C. *J. Chem. Phys.* **2007**, *126*, 244105.

[78] Bytautas, L.; Ruedenberg, K. *J. Chem. Phys.* **2005**, *122*, 154110.

[79] Bytautas, L.; Matsunaga, N.; Nagata, T.; Gordon, M. S.; Ruedenberg, K. *J. Chem. Phys.* **2007**, *127*, 204313.

[80] Bytautas, L.; Matsunaga, N.; Nagata, T.; Gordon, M. S.; Ruedenberg, K. *J. Chem. Phys.* **2007**, *127*, 204301.

[81] Bytautas, L.; Matsunaga, N.; Ruedenberg, K. *J. Chem. Phys.* **2010**, *132*, 074307.

[82] Bytautas, L.; Matsunaga, N.; Scuseria, G. E.; Ruedenberg, K. *J. Phys. Chem. A* **2012**, *116*, 1717.

[83] Peterson, K. A.; Dunning, T. H., Jr. *J. Phys. Chem.* **1995**, *99*, 3898.
[84] Peterson, K. A.; Woon, D. E.; Dunning, T. H., Jr. *J. Chem. Phys.* **1994**, *100*, 7410.
[85] Dunning, T. H., Jr. *J. Chem. Phys.* **1989**, *90*, 1007.
[86] Peterson, K. A.; Wilson, A. K.; Woon, D. E.; Dunning, T. H. *Theor. Chem. Acc.* **1997**, *97*, 251.
[87] Peterson, K. A.; Dunning, T. H., Jr. *J. Mol. Struct. THEOCHEM* **1997**, *400*, 93.
[88] Woon, D. E.; Dunning, J. T. H. *J. Chem. Phys.* **1995**, *103*, 4572.
[89] Dunning, T. H., Jr.; Peterson, K. A.; Wilson, A. K. *J. Chem. Phys.* **2001**, *114*, 9244.
[90] de Jong, W. A.; Harrison, R. J.; Dixon, D. A. *J. Chem. Phys.* **2001**, *114*, 48.
[91] Sendt, K.; Bacskay, G. B. *J. Chem. Phys.* **2000**, *112*, 2227.
[92] Bytautas, L.; Ruedenberg, K. *J. Chem. Phys.* **2004**, *121*, 10919.
[93] Bytautas, L.; Ruedenberg, K. *J. Chem. Phys.* **2006**, *124*, 174304.
[94] Bytautas, L.; Nagata, T.; Gordon, M. S.; Ruedenberg, K. *J. Chem. Phys.* **2007**, *127*, 164317.
[95] Bytautas, L.; Ruedenberg, K. *J. Chem. Phys.* **2010**, *132*, 074109.
[96] Gdanitz, R.; Ahlrichs, R. *Chem. Phys. Lett.* **1988**, *143*, 413.
[97] Szalay, P. G.; Bartlett, R. J. *Chem. Phys. Lett.* **1993**, *214*, 481.
[98] Szalay, PtG.; Müller, T.; Gidofalvi, G.; Lischka, H.; Shepard, R. *Chem. Rev.* **2012**, *112*, 108.
[99] Langhoff, S. R.; Davidson, E. R. *Int. J. Quantum Chem.* **1974**, *8*, 61.
[100] Khait, Y. G.; Jiang, W.; Hoffmann, M. R. *Chem. Phys. Lett.* **2010**, *493*, 1.
[101] Kendall, R. A.; Dunning, T. H., Jr.; Harrison, R. J. *J. Chem. Phys.* **1992**, *96*, 6796.
[102] Xantheas, S. S.; Dunning, T. H. *J. Chem. Phys.* **1993**, *97*, 18.
[103] Feller, D. *J. Chem. Phys.* **1992**, *96*, 6104.
[104] Feller, D. *J. Chem. Phys.* **1993**, *98*, 7059.
[105] Halkier, A.; Helgaker, T.; Jørgensen, P.; Klopper, W.; Olsen, J. *Chem. Phys. Lett.* **1999**, *302*, 437.
[106] Peterson, K. A.; Woon, D. E.; Dunning, J. T. H. *J. Chem. Phys.* **1994**, *100*, 7410.
[107] Schwartz, C. *Phys. Rev. A* **1962**, *126*, 1015.
[108] Schwartz, C. In *Methods in Computational Physics*; Alder, B. J., Fernbach, S., Rotenberg, M., Eds.; Academic: New York, 1963; p 262.
[109] Buenker, R. J. *Chem. Phys. Lett.* **1981**, *72*, 278.
[110] Kallay, M.; Szalay, P. G.; Surjan, P. R. *J. Chem. Phys.* **2002**, *117*, 980.
[111] Lyakh, D. I.; Ivanov, V. V.; Adamowicz, L. *J. Chem. Phys.* **2008**, *128*, 074101.
[112] Ivanov, V. V.; Lyakh, D. I.; Adamowicz, L. *Phys. Chem. Chem. Phys.* **2009**, *11*, 2355.
[113] Evangelista, F. A.; Simmonett, A. C.; Allen, W. D.; Schaefer, H. F.; Gauss, J. *J. Chem. Phys.* **2008**, *128*, 124104.
[114] Andersson, K. *Theor. Chem. Acc.* **1995**, *91*, 31.
[115] Werner, H.J.; Knowles, P. J.; Lindh, R.; Manby, F. R.; Schutz, M.; Celani, P.; Korona, T.; Mitrushenkov, A.; Rauhut, G.; Adler, T. B.; Amos, R. D.; Bernhardsson, A.; Berning, A.; Cooper, D. L.; Deegan, M. J. O; Dobbyn A. J; Eckert, F.; Goll, E.; Hampel, C.; Hetzer, G.; Hrenar, T.; Knizia, G.; Koppl, C.; Liu, Y.; Lloyd, A. W.; Mata, R. A.; May, A. J.; McNicholas, S. J; Meyer, W.; Mura, M. E.; Nicklass, A.; Palmieri, P.; Pfluger, K.; Pitzer, R.; Reiher, M; Schumann, U.; Stoll, H.; Stone, A. J.; Tarroni, R.; Thorsteinsson, T.; Wang, M.; Wolf, A., *MOLPRO A Package of Ab Initio Programs, Version* **2009**, 1.
[116] Celani, P.; Werner, H.-J. *J. Chem. Phys.* **2009**, *112*, 5546.
[117] Angeli, C.; Cimiraglia, R.; Evangelisti, S.; Leininger, T.; Malrieu, J. P. *J. Chem. Phys.* **2001**, *114*, 10252.
[118] Angeli, C.; Borini, S.; Cestari, M.; Cimiraglia, R. *J. Chem. Phys.* **2004**, *121*, 4043.
[119] Angeli, C.; Pastore, M.; Cimiraglia, R. *Theor. Chem. Acc.* **2007**, *117*, 743.
[120] Hirao, K. *Chem. Phys. Lett.* **1993**, *201*, 59.

[121] Hirao, K. *Chem. Phys. Lett.* **1992**, *190*, 374.
[122] Khait, Y. G.; Song, J.; Hoffmann, M. R. *J. Chem. Phys.* **2002**, *117*, 4133.
[123] Jiang, W.; Khait, Y. G.; Hoffmann, M. R. *J. Phys. Chem. A* **2009**, *113*, 4374.
[124] Müller, T.; Dallos, M.; Lischka, H.; Dubrovay, Z.; Szalay, P. G. *Theor. Chem. Acc.* **2001**, *105*, 227.
[125] Pople, J. A.; Seeger, R.; Krishnan, R. *Int. J. Quantum Chem. Symp.* **1977**, *11*, 149.
[126] Meissner, L. *Chem. Phys. Lett.* **1988**, *146*, 204.
[127] Lee, T. J. *Chem. Phys. Lett.* **2003**, *372*, 362.
[128] Douglas, J. E.; Rabinovitch, B. S.; Looney, F. S. *J. Chem. Phys.* **1955**, *23*, 315.
[129] Olbrich, G.; Potzinger, P.; Reimann, B.; Walsh, R. *Organometallics* **1984**, *3*, 1267.
[130] Frisch, M. J.; Trucks, G. W.; Schlegel, H. B.; Scuseria, G. E.; Robb, M. A.; Cheeseman, J. R.; Scalmani, G.; Barone, V.; Mennucci, B.; Petersson G. A.; Nakatsuji, H.; Caricato, M.; Li, X.; Hratchian, H. P.; Izmaylov A. F.; Bloino, J.; Zheng, G.; Sonnenberg, J. L.; Hada, M.; Ehara, M.; Toyota, K.; Fukuda, R.; Hasegawa, J.; Ishida, M.; Nakajima, T.; Honda, Y.; Kitao, O.; Nakai, H.; Vreven, T.; Montgomery, J. J. A.; Peralta, J. E; Ogliaro, F.; Bearpark, M.; Heyd, J. J.; Brothers, E.; Kudin, K. N.; Staroverov, V. N.; Kobayashi, R.; Normand, J.; Raghavachari, K.; Rendell, A.; Burant, J. C.; Iyengar, S. S.; Tomasi, J.; Cossi, M.; Rega, N.; Millam, J. M.; Klene, M.; Knox J. E.; Cross, J.B.; Bakken, V.; Adamo, C.; Jaramillo, J.; Gomperts, R.; Stratmann, R.E; Yazyev, O; Austin, A. J.; Cammi, R.; Pomelli, C.; Ochterski J. W.; Martin, R. L.; Morokuma, K.; Zakrzewski, V. G.; Voth, G. A.; Salvador, P.; Dannenberg, J.J.; Dapprich, S.; Daniels, A. D.; Farkas, Ö.; Foresman, J. B.; Ortiz, J. V.; Cioslowski, J.; Fox, D. J. *Revision A.* **2009**, *1*.
[131] Valiev, M.; Bylaska, E. J.; Govind, N.; Kowalski, K.; Straatsma, T. P.; Van, D. H. J. J.; Wang, D.; Nieplocha, J.; Apra, E.; Windus, T. L.; de, J. W. A. *Comput. Phys. Commun.* **2009**, *181*, 1477.

CHAPTER THREE

On the Orthogonality of Orbitals in Subsystem Kohn–Sham Density Functional Theory

Yuriy G. Khait and Mark R. Hoffmann[1]
Chemistry Department, University of North Dakota, Grand Forks, North Dakota, USA
[1]Corresponding author: E-mail: mhoffmann@chem.und.edu

Contents

Abstract

A rigorous analysis of the requirements on orbitals of subsystems in the context of Kohn–Sham density functional theory is presented. It is found that conventional constrained electron density formulations, which neglect explicit consideration of orthogonality requirements between subsystems, are exactly correct only in the limit of infinitely separated subsystems. A new method is suggested that takes into account strictly orthogonality constraints and perspectives for its practical implementation are discussed. The performed analysis also provides a practical method for including corrections to conventional methods that approximately compensate for non-orthogonality. It is also shown that the same considerations apply to wave function-in-DFT embedding schemes.

1. INTRODUCTION

The subsystem formulation of density functional theory (SDFT), where each subsystem and its interactions are described at the Kohn–Sham (KS) DFT level [1], was originally suggested for studying the properties of

53

condensed matter consisting of closed-shell atoms or ions [2,3,4], chemical processes in solutions [5], and weakly bound molecular complexes [6]. In this formulation, which is also referred to as the DFT-in-DFT embedding model, it is supposed that: (1) a system can be divided into two (or more) interacting subsystems with integer numbers of electrons, (2) the entire density can be partitioned into the sum of the subsystems' densities, and (3) the subsystems' KS orbitals are not required to be orthogonal to those from different subsystem [7]. Minimization of the total energy functional under such conditions leads to a system of coupled KS-like equations for the subsystems' orbitals [2,3,4,5,6], which are referred to as the KS equations with constrained electron density (KSCED) [6,7]. These equations involve, in addition to the conventional KS potential related to the exchange-correlation functional ($E_{xc}[\rho]$), derivatives of the noninteracting kinetic energy functional ($T_s[\rho]$) [1,8]. The calculations based on the KSCED equations, which are solved for one subsystem at a time while keeping the density of the other one fixed, are often referred to as orbital-free calculations since the effect of subsystems other than the one that is being optimized is completely determined by their currently fixed density.

Based on the same ideology as the DFT-in-DFT model, a wave function theory-in-DFT (WFT-in-DFT) embedding approach was also developed [9,10]. In this approach, a subsystem of primary interest (i.e. the embedded subsystem) is described using an *ab initio* wave function-based method, while the remaining part (i.e. the environment) and the interactions between the subsystems are described at the KS DFT level. The main conceptual advantage of the WFT-in-DFT approach is that this approach not only allows one to describe a potentially complex multiconfigurational structure for the ground state of the embedded subsystem but also ensures a systematic description of excited states of the system if such states are mainly related to excitations within the embedded subsystem (see refs. 11,12,13 and references therein).

There are three main KSCED-based methods. In KSCED(s) calculations [6], orbitals of each subsystem are expanded over atomic functions centered on all nuclei of the whole system (a so-called "supermolecular" atomic basis); in contrast, in performing KSCED(m) calculations, one expands orbitals of each subsystem only over atomic functions centered on its own nuclei ("monomer" atomic basis sets [7]). A particular case of the KSCED(m) method is the frozen density embedding (FDE) approximation [5,7], where the density of the environment is kept unchanged (i.e. frozen). These KSCED-based methods and, especially, the FDE approximation proved to be quite successful, useful, and efficient in many practical applications: in modeling processes in solutions [14,15], for the description of solvent effects on absorption spectra [16,17,18], electron spin resonance parameters [19], and nuclear magnetic resonance chemical shifts [20]; in modeling charge transfer reactions [21]; and in calculating interaction energies of weak intermolecular complexes and subsystems connected by hydrogen bonds (see refs. [7,22,23,24] and references therein), etc.

These KSCED methods, however, break down if interactions between subsystems involve π-electrons [25,26] or have a non-negligible covalent character (see refs. [26,27,28] and references therein) and proved to be unreliable for predicting geometries of even noncovalently bound inter-molecular complexes, especially in the case of using generalized gradient approximation-type (GGA) functionals instead of local density approximation (LDA) functionals. For example, for all 21 intermolecular complexes considered in ref. 29, KSCED GGA calculations worsen the KSCED LDA geometries appreciably (N.B. the errors in the intermolecular distances are increased two-to-three times and can reach 0.5 Å), while conventional KS GGA calculations improve the KS LDA geometries significantly and systematically, and the final errors do not exceed 0.1 Å [29]. As a rule, these failures of the KSCED-based methods are supposed to result from an inaccuracy of the available approximations to derivatives of $T_s[\rho]$, assuming that, for the exact $T_s[\rho]$, the basic KSCED equations are equivalent to the KS equations for the whole system and, hence, have the same range of applicability including covalently bonded subsystems [6,7,26,28,30,31].

The main goal of the present work is to show that the reason for failures of the KSCED-based methods can also be related to an inaccuracy of the basic KSCED equations themselves in regions where the overlaps of subsystems' densities are non-negligible. We show that the total density is representable as the sum of densities of subsystems if and only if orbitals of different subsystems are orthogonal and, as a consequence, that a basic assumption of the orbital SDFT, i.e. that the subsystems' KS orbitals are not required to be orthogonal to those from different subsystems [7], restricts the applicability of the basic KSCED equations to the case of weakly over-lapping (strictly speaking, nonoverlapping) subsystem densities. In the region where these densities overlap strongly, their sum can differ significantly from the accurate total density for given nonorthogonal orbitals of the subsystems. As a consequence, disregarding the nonorthogonality of orbitals in such a region can lead, in the course of calculations, to using an incorrect total density and gradient and, hence, to an incorrect evaluation of the nonadditive kinetic energy functional derivative for any (even for the exact) functional $T_s[\rho]$. From this viewpoint, speculations [27] about the correctness of the "conjointness" hypothesis of Lee et al. [32] about the structure of the $T_s[\rho]$ functional, based on calculations performed with an incorrect total density, seem to be quite questionable.

In the present work, new equations for the subsystems' KS orbitals are obtained, which additionally take into account the condition of their "external" (i.e. intermolecular) orthogonality. For the exact kinetic energy functional, these equations are completely equivalent to the KS equations for the whole system at any separation of subsystems and, hence, they are also applicable even in the case of strongly overlapping densities. We show that subsystems orbitals that satisfy the new equations are related to the supermolecular KS orbitals through a simple unitary (orthogonal) transformation. Furthermore, and potentially noteworthy, knowledge of correct subsystem orbitals is enough to reproduce uniquely both the exact KS orbitals of the whole system and their energies. Practical implementation of the obtained equations does not seem to be essentially more difficult than that of the KSCED equations. In particular, a possible alternative to the KSCED(m) method is discussed. A simple way to correct the sum of densities, by taking into account their nonorthogonality, in order to obtain a more accurate total density and to use it in conventional KSCED calculations is also suggested. Using this modified density should extend the applicability of conventional KSCED to cases of non-negligible overlaps of densities.

It will be shown that the WFT-in-DFT approach [9,10] can also break down in the region where the subsystems' densities overlap strongly if one does not impose the restriction of external orthogonality on the orbitals of the subsystems. A way to take this requirement into account in optimizing the embedded orbitals within wave function-based methods is suggested and discussed in detail.

2. THEORY

2.1. KS Subsystem Equations

The ground state density, $\rho(\mathbf{r})$, of a system of N interacting electrons moving in a field $v(\mathbf{r})$ of nuclei is determined by the KS equation, which in its Euler form [1] is

$$\frac{\delta T_s[\rho]}{\delta \rho(\mathbf{r})} + v_s([\rho]; \mathbf{r}) = \mu_s, \tag{3.1}$$

where

$$v_s([\rho]; \mathbf{r}) = v(\mathbf{r}) + \int \frac{\rho(\mathbf{r}\prime)}{|\mathbf{r} - \mathbf{r}\prime|} d\mathbf{r}\prime + \frac{\delta E_{xc}[\rho]}{\delta \rho(\mathbf{r})} \qquad (3.2)$$

is the KS one-electron effective potential. The $\rho(\mathbf{r})$ from Eqn (3.1) is representable through orthonormal KS orbitals $\{\varphi_m^{KS}\}_{m=1}^{N}$,

$$\rho(\mathbf{r}) = \sum_{m=1}^{N} |\varphi_m^{KS}(\mathbf{r})|^2, \qquad (3.3)$$

if these orbitals satisfy the equations

$$h^{KS}|\varphi_m^{KS}\rangle = \varepsilon_m^{KS}|\varphi_m^{KS}\rangle, \quad (m \in [1, N]), \qquad (3.4)$$

where

$$h^{KS}(\mathbf{r}) = -\frac{1}{2}\nabla^2 + v_s([\rho]; \mathbf{r}) \qquad (3.5)$$

Hereafter, for simplicity of notation, we do not write equations for α- and β-spin orbitals separately.

If one supposes that the system can be divided into two interacting subsystems with integer numbers N_A and $N_B = N - N_A$ of electrons, associated with well localized sets A and B of nuclei, and that the entire density $\rho(\mathbf{r})$ can be partitioned into the sum of N_A- and N_B-representable [33] densities of the subsystems,

$$\rho(\mathbf{r}) = \rho_A(\mathbf{r}) + \rho_B(\mathbf{r}), \qquad (3.6)$$

then, in analogy with Eqn (3.1), the stationary conditions for the total energy functional relative to variations of $\rho_A(\mathbf{r})$ and $\rho_B(\mathbf{r})$ lead to the coupled Euler equations

$$\frac{\delta T_s[\rho_A]}{\delta \rho_A(\mathbf{r})} + v_{eff}^A([\rho, \rho_A]; \mathbf{r}) = \mu_s^A, \qquad (3.7a)$$

$$\frac{\delta T_s[\rho_B]}{\delta \rho_B(\mathbf{r})} + v_{eff}^B([\rho, \rho_B]; \mathbf{r}) = \mu_s^B \qquad (3.7b)$$

These equations involve the effective potentials [2,3,4,5],

$$v_{eff}^I([\rho, \rho_I]; \mathbf{r}) = v_s([\rho]; \mathbf{r}) + v_T^I([\rho, \rho_I]; \mathbf{r}), \quad (I = A, B) \qquad (3.8)$$

each of which differs from the supermolecular KS potential $v_s([\rho];\mathbf{r})$ from Eqn (3.2) only by the subsystem specific "kinetic potential" [2] due to the nonadditive nature of the functional $T_s[\rho]$,

$$v_T^I([\rho, \rho_I]; \mathbf{r}) = \frac{\delta T_s[\rho]}{\delta \rho(\mathbf{r})} - \frac{\delta T_s[\rho_I]}{\delta \rho_I(\mathbf{r})} \tag{3.9}$$

For the self-consistent solution (ρ_A, ρ_B) of Eqn (3.7), each of these equations can be seen to lead to Eqn (3.1). Thus, Eqn (3.7a) and (3.7b) are consistent and determine the total density $\rho(\mathbf{r})$ from Eqn (3.1) if and only if $\mu_s^A = \mu_s^B = \mu_s$ [34]. Hence, only under such circumstances is $\rho(\mathbf{r})$ representable as the sum $\rho_A(\mathbf{r}) + \rho_B(\mathbf{r})$.

2.2. Orthogonality Conditions

In reformulating Eqn (3.7) in terms of the subsystems' KS orbitals, there also must exist conditions on these orbitals such that $\rho(\mathbf{r})$ from Eqn (3.3) is representable as the sum of the subsystems' densities. In principle, there are a few directions that one could pursue in achieving this goal; one alternative is sketched at the end of Section 3.2.4, building on developments in Sections 3.2.3 and 3.2.4. The question of alternatives is revisited in Section 3.3. The most direct, and arguably practical, solution is the one presented here. Within a potentially exact SDFT, knowledge of correct sets of the subsystems' KS orbitals, $|\overline{\varphi}^A\rangle = |\varphi_1^A, ..., \varphi_{N_A}^A\rangle$ and $|\overline{\varphi}^B\rangle = |\varphi_1^B, ..., \varphi_{N_B}^B\rangle$, each of which is supposed to be internally orthonormal and determines the corresponding subsystems' density,

$$\rho_A(\mathbf{r}) = \sum_{a=1}^{N_A} |\varphi_a^A(\mathbf{r})|^2, \quad \rho_B(\mathbf{r}) = \sum_{b=1}^{N_B} |\varphi_b^B(\mathbf{r})|^2, \tag{3.10}$$

should be enough to reproduce completely the exact KS orbitals $\{\varphi_m^{KS}\}_{m=1}^N$ of the whole system. This means that the orthonormal orbitals $\{\varphi_m^{KS}\}_{m=1}^N$ can be expanded in the basis of the subsystems' orbitals and, hence, span the same orbital subspace, $L_{\overline{\varphi}}$, as does the composite set $|\overline{\varphi}\rangle = |\overline{\varphi}^A, \overline{\varphi}^B\rangle$. Since the total density $\rho(\mathbf{r})$ from Eqn (3.3) is invariant under any unitary transformation of the orbitals $\{\varphi_m^{KS}\}_{m=1}^N$, $\rho(\mathbf{r})$ depends only on the subspace $L_{\overline{\varphi}}$ and not on the choice of an orthonormal basis set within this subspace. The key question is "Under what conditions on the sets $\overline{\varphi}^A$ and $\overline{\varphi}^B$ is the sum $\rho_A(\mathbf{r}) + \rho_B(\mathbf{r})$ representable as the diagonal quadratic form Eqn (3.3) over an orthonormal orbital set $|\overline{\varphi}^{orth}\rangle = |\varphi_1^{orth}, ..., \varphi_N^{orth}\rangle$ within $L_{\overline{\varphi}}$?". The answer is almost obvious: the sets $\overline{\varphi}^A$ and $\overline{\varphi}^B$ must be orthogonal. Indeed, let us define an orthonormal set $\overline{\varphi}^{\text{orth}}$ within $L_{\overline{\varphi}}$, e.g. as $|\overline{\varphi}^{\text{orth}}\rangle = |\overline{\varphi}\rangle \mathbf{S}^{-\frac{1}{2}}$, where $\mathbf{S}$ is the overlap matrix

$$
\mathbf{S} = \langle \overline{\varphi} | \overline{\varphi} \rangle = \begin{pmatrix} \mathbf{I}_{AA} & \mathbf{S}_{AB} \\ \mathbf{S}_{BA} & \mathbf{I}_{BB} \end{pmatrix}. \tag{3.11}
$$

Then, based on Eqn (3.10), one has

$$
\rho_A(\mathbf{r}) + \rho_B(\mathbf{r}) = \sum_{k,l=1}^{N} \varphi_k^{\text{orth}}(\mathbf{r}) S_{kl} \varphi_l^{\text{orth}*}(\mathbf{r})
$$

$$
= \sum_{k=1}^{N} |\varphi_k^{\text{orth}}(\mathbf{r})|^2 + \sum_{k,l=1}^{N} \varphi_k^{\text{orth}}(\mathbf{r})(\mathbf{S}-\mathbf{I})_{kl} \varphi_l^{\text{orth}*}(\mathbf{r}),
$$

$$
\tag{3.12}
$$

and, hence, the sum $\rho_A(\mathbf{r}) + \rho_B(\mathbf{r})$ for any given $\overline{\varphi}^A$ and $\overline{\varphi}^B$ is representable as the diagonal quadratic form Eqn (3.3) within $L_{\overline{\varphi}}$ if and only if $\mathbf{S} = \mathbf{I}$, i.e. if the composite set $|\overline{\varphi}\rangle = |\overline{\varphi}^A, \overline{\varphi}^B\rangle$ is orthonormal. This means that the ground state density in its KS orbital form Eqn (3.3) can be represented as the sum of the subsystems' densities Eqn (3.10) only if the subsystems' KS orbitals are both internally orthonormal, i.e. $\langle \varphi_a^A | \varphi_{a'}^A \rangle = \delta_{aa'}$ and $\langle \varphi_b^B | \varphi_{b'}^B \rangle = \delta_{bb'}$, and externally orthogonal, i.e. $\langle \varphi_a^A | \varphi_b^B \rangle = 0$. If the last condition is not satisfied, then the sum $\rho_A(\mathbf{r}) + \rho_B(\mathbf{r})$ must additionally be corrected for given sets $\overline{\varphi}^A$ and $\overline{\varphi}^B$ by taking into account their overlap to lead to the correct total density, defined as [35]

$$
\rho(\mathbf{r}) = \sum_{k=1}^{N} |\varphi_k^{\text{orth}}(\mathbf{r})|^2 = \sum_{c,d=1}^{N_A+N_B} \varphi_c(\mathbf{r})(\mathbf{S}^{-1})_{cd} \varphi_d^*(\mathbf{r}) \tag{3.13}
$$

This observation is especially important in the case of the KSCED equations [2,3,4,5,7], which do not impose the condition of external orthogonality on the subsystems' orbitals (although they do enforce internal orthonormality) and could lead to a total density [calculated as the sum $\rho_A(\mathbf{r}) + \rho_B(\mathbf{r})$] that differs significantly from the exact KS density Eqn (3.3) in the region where $\overline{\varphi}^A$ and $\overline{\varphi}^B$ are strongly overlapping. Consequently, even if the exact functional $T_s[\rho]$ were known, the KSCED equations would not be equivalent to the KS Eqn (3.4) for the whole system (except in the limiting case of infinite separation of subsystems). One can try to overcome this problem of the KSCED scheme, at least partially, by correcting the sum $\rho_A(\mathbf{r}) + \rho_B(\mathbf{r})$, obtained in KSCED calculations, by taking into account the overlap of the nonorthogonal sets $\overline{\varphi}^A$ and $\overline{\varphi}^B$ at each iteration. A simple way of performing such corrections is considered in Section 3.2.4, after an exact treatment of nonorthogonality presented in the following section.

2.3. A Modified KSCED Scheme

Equations for the KS orbitals of subsystems that respect their orthogonality can be obtained in the same way as the KSCED equations have been obtained [2,3,4,5,7], but with the additional condition that the KS electronic energy density functional for the whole interacting system [7,36]

$$E^S[\overline{\varphi}^A, \overline{\varphi}^B] = \sum_{a=1}^{N_A} \langle \varphi_a^A | -\frac{1}{2}\nabla^2 | \varphi_a^A \rangle + \sum_{b=1}^{N_B} \langle \varphi_b^B | -\frac{1}{2}\nabla^2 | \varphi_b^B \rangle + V[\rho] + J[\rho]$$

$$+ E_{xc}[\rho] + T_s[\rho] - T_s[\rho_A] - T_s[\rho_B] \tag{3.14}$$

must be stationary [37] with respect to orbital variations, while enforcing not only internal orthonormality but also external orthogonality of orbitals. This leads to the consideration of the Lagrangian

$$\Omega[\overline{\varphi}^A, \overline{\varphi}^B] = E^S[\overline{\varphi}^A, \overline{\varphi}^B] - \sum_{I=A,B} \sum_{c,c'=1}^{N_I} \Theta_{c'c}^I \langle \varphi_c^I | \varphi_{c'}^I \rangle$$

$$- \sum_{a=1}^{N_A} \sum_{b=1}^{N_B} \alpha_{ba} \langle \varphi_a^A | \varphi_b^B \rangle - \sum_{a=1}^{N_A} \sum_{b=1}^{N_B} \beta_{ab} \langle \varphi_b^B | \varphi_a^A \rangle; \tag{3.15}$$

the stationary conditions of Eqn (3.15) with respect to independent variations of orbitals, e.g. of subsystem A, lead to the equations

$$[h^{KS} + v_T^A] | \varphi_a^A \rangle = \sum_{a'=1}^{N_A} | \varphi_{a'}^A \rangle \Theta_{a'a}^A + \sum_{b=1}^{N_B} | \varphi_b^B \rangle \alpha_{ba} \quad (a \in [1, N_A]) \tag{3.16}$$

Since, based on Eqn (3.16), the matrix $\|\Theta_{a'a}^A\|$ is symmetrical, and the total energy (Eqn (3.14)) is invariant under any unitary transformation within $\overline{\varphi}^A$, one can use a set of the canonical orbitals for which $\Theta_{aa'}^A = \delta_{aa'}\varepsilon_a^A$ and rewrite Eqn (3.16) as

$$[h^{KS} + v_T^A] | \varphi_a^A \rangle = | \varphi_a^A \rangle \varepsilon_a^A + \sum_{b=1}^{N_B} | \varphi_b^B \rangle \alpha_{ba} \tag{3.17}$$

Based on Eqn (3.17), one gets

$$\alpha_{ba} = \alpha_{ab} = \langle \varphi_b^B | h^{KS} + v_T^A | \varphi_a^A \rangle, \tag{3.18}$$

and, hence, this equation can be rewritten as the eigenvalue problem

$$(1 - P^B)[h^{KS} + v_T^A]|\varphi_a^A\rangle = |\varphi_a^A\rangle \varepsilon_a^A, \tag{3.19}$$

where $P^B = |\overline{\varphi}^B\rangle\langle\overline{\varphi}^B| = \sum_{b=1}^{N_B} |\varphi_b^B\rangle\langle\varphi_b^B|$ is the projector on the KS orbitals of subsystem B. Equation (3.19) can easily be verified to guarantee that $\overline{\varphi}^A$ will be orthogonal to a given set $\overline{\varphi}^B$. Since $(1 - P^B)|\overline{\varphi}^A\rangle = |\overline{\varphi}^A\rangle$, Eqn (3.19) can finally be rewritten as an eigenvalue problem with a reduced, Hermitian, one-electron Hamiltonian

$$(1 - P^B)[h^{KS} + v_T^A](1 - P^B)|\varphi_a^A\rangle = |\varphi_a^A\rangle \varepsilon_a^A \tag{3.20a}$$

In the same way, one obtains the equations for the KS orbitals of subsystem B, orthogonal to $\overline{\varphi}^A$,

$$(1 - P^A)[h^{KS} + v_T^B](1 - P^A)|\varphi_b^B\rangle = |\varphi_b^B\rangle \varepsilon_b^B, \tag{3.20b}$$

where $P^A = |\overline{\varphi}^A\rangle\langle\overline{\varphi}^A| = \sum_{a=1}^{N_A} |\varphi_a^A\rangle\langle\varphi_a^A|$. Note that the Lagrange multipliers

$$\beta_{ab} = \beta_{ab} = \langle \varphi_a^A | h^{KS} + v_T^B | \varphi_b^B \rangle, \tag{3.21}$$

which appear in deriving Eqn (3.20b), are not equal to those from Eqn (3.18).

It should be emphasized that in contrast to the KSCED equations, the new Eqn (3.20) require the subsystems' orbitals be eigenvectors not of the KSCED Hamiltonians $(h^{KS} + v_T^I)$ $(I = A, B)$ but of the reduced Hamiltonians, which involve additionally the projectors $1 - P^I$. Consequently, in contrast to the KSCED equations, the new Eqn (3.20) take into account explicitly interactions between orbitals of different subsystems through the blocks $\langle\overline{\varphi}^B | h^{KS} + v_T^I | \overline{\varphi}^A\rangle$ [see Eqn (3.18) and (3.21)]. Since Eqn (3.20) consider the same (more exactly, equivalent [7,34]) total energy functional as does Eqn (3.4) and take into account exactly all the same restrictions on the orbitals, these new equations are completely equivalent for exact $T_s[\rho]$ to the KS Eqn (3.4) if both Eqns (3.20) and (3.4) are projected (from the left) on the same (supermolecular) set of atomic basis functions. This means that the self-consistent sets $|\overline{\varphi}^{KS}\rangle$ and $|\overline{\varphi}\rangle = |\overline{\varphi}^A, \overline{\varphi}^B\rangle$ determine the same total density and, hence, are related through a unitary transformation,

$$|\overline{\varphi}^{KS}\rangle = |\overline{\varphi}\rangle \mathbf{U} = |\overline{\varphi}^A, \overline{\varphi}^B\rangle \mathbf{U} \quad (\mathbf{U}^+ = \mathbf{U}^{-1}), \tag{3.22}$$

i.e. these sets are two orthonormal bases within the same N-dimensional orbital space. Based on Eqns (3.4) and (3.22), the required unitary transformation ($\mathbf{U}$) is determined by the condition that the matrix

$$\boldsymbol{\varepsilon}^{KS} = \langle \overline{\varphi}^{KS} | h^{KS} | \overline{\varphi}^{KS} \rangle = \mathbf{U}^+ \langle \overline{\varphi} | h^{KS} | \overline{\varphi} \rangle \mathbf{U} \tag{3.23}$$

must be diagonal, i.e. $\mathbf{U}$ must diagonalize the KS Hamiltonian matrix, which is Hermitian, in the basis of the subsystems' orbitals,

$$\langle \overline{\varphi} | h^{KS} | \overline{\varphi} \rangle = \begin{pmatrix} \mathbf{h}_{AA}^{KS} & \mathbf{h}_{AB}^{KS} \\ \mathbf{h}_{BA}^{KS} & \mathbf{h}_{BB}^{KS} \end{pmatrix} \tag{3.24}$$

Thus, as has been expected, knowledge of the self-consistent KS orbitals of the subsystems is enough to reproduce completely both the KS orbitals of the whole system and their energies. Note that, Eqns (3.23) and (3.24) can be used in practice as a simple and reliable test of the accuracy of kinetic energy functionals: in the case that $T_s[\rho]$ is exact, eigenvalues of matrix Eqn (3.24) constructed with $\overline{\varphi}^A$ and $\overline{\varphi}^B$ from Eqn (3.20) must be equal to the KS one-electron energies from Eqn (3.4) *for any separation of subsystems*.

Practical implementation of new Eqn (3.20) does not seem to be essentially more difficult than that of the KSCED equations since all additional calculations related to projectors P^I lead to calculations of a subsystem density matrix in the basis of atomic functions. However, for ensuring external orthogonality, orbitals of each subsystem must be expanded not only over its own (monomer) atomic basis set but also over its atomic functions of the complementary subsystem. If one performs calculations based on Eqn (3.20) and uses the supermolecular atomic basis set for each subsystem, one obtains a direct analog to the KSCED(s) method. On the other hand, if one extends the monomer atomic basis set of each subsystem with atomic functions (or, at least, some of the functions) centered on a few common nuclei important for describing the overlap region and projects Eqn (3.20) only on such subsystem specific "extended monomer" basis sets, one obtains a method that is the analog of the KSCED(m) method [7] but is potentially applicable for studying even strongly bonded subsystems. The development of an alternative to the FED approximation that would be based on the new equations and would use the subsystem specific extended monomer basis sets also seems to be possible but this task requires additional analysis.

2.4. Approximate Inclusion of Orthogonality Constraints in KSCED

As mentioned above the fact that the KSCED scheme does not take into account the external orthogonality condition leads to the following serious consequence: in each iteration, one calculates the potentials $v_s([\rho];$ $\mathbf{r})$ and $v_T^I[(\rho, \rho_I); \mathbf{r}]$ using an incorrect total density $[\rho_A(\mathbf{r}) + \rho_B(\mathbf{r})]$, which can differ from the correct one from Eqn (3.13) significantly in the region where the overlap of $\overline{\varphi}^A$ and $\overline{\varphi}^B$ is non-negligible. One should be able to overcome this problem, at least partially, if one uses, instead of the sum of densities, a density that would be close to the accurate one from Eqn (3.13). With this aim, and representing the overlap matrix from Eqn (3.11) as $\mathbf{S} = \mathbf{I} + \boldsymbol{\Delta}$, where $\boldsymbol{\Delta} = \begin{pmatrix} \mathbf{0}_{AA} & \mathbf{S}_{AB} \\ \mathbf{S}_{BA} & \mathbf{0}_{BB} \end{pmatrix}$, one obtains the expansion

$$\mathbf{S}^{-1} = (\mathbf{I} + \boldsymbol{\Delta})^{-1} = \mathbf{I} - \boldsymbol{\Delta} + \boldsymbol{\Delta}^2 - \boldsymbol{\Delta}^3 + \boldsymbol{\Delta}^4 - \cdots$$

$$= \mathbf{I} - \boldsymbol{\Delta} + \boldsymbol{\Delta}^2 - \boldsymbol{\Delta}^3 + \boldsymbol{\Delta}^4 - \boldsymbol{\Delta}^5(\mathbf{I} + \boldsymbol{\Delta})^{-1} \tag{3.25}$$

Taking into account that

$$(\boldsymbol{\Delta}^{2n})_{AB} = 0, \quad (\boldsymbol{\Delta}^{2n})_{BA} = 0, \quad (\boldsymbol{\Delta}^{2n+1})_{AA} = 0,$$

$$(\boldsymbol{\Delta}^{2n+1})_{BB} = 0, \quad (n = 0, 1, 2, \ldots) \tag{3.26}$$

the matrix $\mathbf{S}^{-1}$ can be rewritten in the following simple form, which is accurate to fourth order in $\mathbf{S}_{AB}$,

$$\mathbf{S}^{-1} \cong \begin{pmatrix} \mathbf{I}_{AA} + (\mathbf{I}_{AA} + \mathbf{d}_{AA})\mathbf{d}_{AA} & -(\mathbf{I}_{AA} + \mathbf{d}_{AA})\mathbf{S}_{AB} \\ -(\mathbf{I}_{BB} + \mathbf{d}_{BB})\mathbf{S}_{BA} & \mathbf{I}_{BB} + (\mathbf{I}_{BB} + \mathbf{d}_{BB})\mathbf{d}_{BB} \end{pmatrix}, \tag{3.27}$$

where $\mathbf{d}_{AA} = (\boldsymbol{\Delta}^2)_{AA} = \mathbf{S}_{AB}\mathbf{S}_{BA}$ and $\mathbf{d}_{BB} = (\boldsymbol{\Delta}^2)_{BB} = \mathbf{S}_{BA}\mathbf{S}_{AB}$. Inserting Eqn (3.27) into Eqn (3.13), the total density associated with given nonorthogonal sets $\overline{\varphi}^A$ and $\overline{\varphi}^B$ will be approximated as follows:

$$\rho(\mathbf{r}) = \rho_A(\mathbf{r}) + \rho_B(\mathbf{r}) + \rho_{AB}(\mathbf{r}) + \Delta\rho_A(\mathbf{r}) + \Delta\rho_B(\mathbf{r}), \tag{3.28}$$

where $\rho_A(\mathbf{r})$ and $\rho_B(\mathbf{r})$ are still determined by Eqn (3.10), and

$$\rho_{AB}(\mathbf{r}) = -2 \sum_{a=1}^{N_A} \sum_{b=1}^{N_B} \varphi_a^A(\mathbf{r})\varphi_b^{B*}(\mathbf{r})[(\mathbf{I}_{AA} + \mathbf{d}_{AA})\mathbf{S}_{AB}]_{ab}, \tag{3.29}$$

$$\Delta\rho_A(\mathbf{r}) = \sum_{a,a'=1}^{N_A} \varphi_a^A(\mathbf{r})\varphi_{a'}^{A*}(\mathbf{r})[(\mathbf{I}_{AA} + \mathbf{d}_{AA})\mathbf{d}_{AA}]_{aa'}, \tag{3.30}$$

$$\Delta\rho_B(\mathbf{r}) = \sum_{b,b'=1}^{N_B} \varphi_b^B(\mathbf{r})\varphi_{b'}^{B*}(\mathbf{r})[(\mathbf{I}_{BB} + \mathbf{d}_{BB})\mathbf{d}_{BB}]_{bb'} \tag{3.31}$$

Note that $\rho_{AB}(\mathbf{r})$, which is related to orbital pairs between subsystems, makes first- and third-order corrections to the total density, while $\Delta\rho_A(\mathbf{r})$ and $\Delta\rho_B(\mathbf{r})$ determine the corrections at second and fourth orders. Thus, $\rho(\mathbf{r})$ from Eqn (3.28) takes into account the nonorthogonality of orbital sets $\overline{\varphi}^A$ and $\overline{\varphi}^B$ up to fourth order in the overlap matrix $\mathbf{S}_{AB}$. It stands to reason that using a more accurate total density instead of the sum $\rho_A(\mathbf{r}) + \rho_B(\mathbf{r})$, in each iteration, should increase the accuracy and reliability of conventional KSCED calculations, especially when one considers a region in which the overlap of the subsystems' densities is non–negligible.

Based on the performed analysis, one could obtain correct equations for the total density using *nonorthogonal* basis sets $\overline{\varphi}^A$ and $\overline{\varphi}^B$. Such a derivation would proceed by considering the stationary conditions of the total energy functional, Eqn (3.14), with the total density determined by Eqn (3.13), with respect to variations of orbitals that are only constrained by conditions of internal orthonormality. Although the resulting equations are sure to be more cumbersome than the formulation for orthogonal orbital sets presented herein, it is conceivable that benefits accrued because of, e.g. improved localization would sufficiently offset the increase in complexity in some circumstances. This constitutes one of the alternatives, alluded to at the beginning of Section 3.2.2, for obtaining subsystem KS equations whose solutions are exactly equivalent to those of the KS equations for the whole system.

2.5. Orthogonality Constraints within the WFT-in-DFT Approach

The WFT–in–DFT approach is characterized by an embedded subsystem (say A) being described using an *ab initio* wave function-based method, while both the environment (B) and the interactions between the subsystems being described at the KS DFT level. Within such an approach, the total density is represented as the sum of the subsystems' densities,

$$\rho(\mathbf{r}) = \rho_A^{\Psi_A}(\mathbf{r}) + \rho_B(\mathbf{r}), \tag{3.32}$$

where $\rho_B(\mathbf{r})$ is still determined by the diagonal quadratic form Eqn (3.10) over the environment KS orbitals $\{\varphi_b^B\}_{b=1}^{N_B}$ from Eqn (3.20b) but the density of the embedded subsystem, described by a normalized wave function Ψ_A, is determined as

$$\rho_A^{\Psi_A}(\mathbf{r}) = \langle \Psi_A | \hat{\rho}_A(\mathbf{r}) | \Psi_A \rangle$$

$$= \sum_{a,a'} \varphi_a^A(\mathbf{r}) \varphi_{a'}^A(\mathbf{r}) \langle \Psi_A | a_{a\alpha}^+ a_{a'\alpha} + a_{a\beta}^+ a_{a'\beta} | \Psi_A \rangle \qquad (3.33)$$

Equation (3.33) involves the creation (a_a^+) and annihilation (a_a) operators associated with a new set of orthonormal embedded orbitals $\{\varphi_a^A\}$ whose number can be significantly larger than the N_A needed for a single determinant. These new orbitals, in contrast to their KS counterparts, are used for constructing the potentially highly correlated embedded state Ψ_A [N.B. Note that spins of orbitals in each orbital pair $\varphi_a^A(\mathbf{r})\varphi_{a'}^A(\mathbf{r})$ in Eqn (3.33) must be the same, i.e. $\alpha\alpha$ or $\beta\beta$ (see, e.g. ref. 38).]. It can be shown that given a set of embedded orbitals $\{\varphi_a^A\}$ and the density of the environment $\rho_B(\mathbf{r})$, the configurational structure of the wave function Ψ_A can be determined from the self-consistent eigenvalue problem [34,39]

$$\left[H^A + \int v_{\text{emb}}^A([\rho_A^{\Psi_A}, \rho_B]; \mathbf{r}) \hat{\rho}_A(\mathbf{r}) \mathrm{d}\mathbf{r}] | \Psi^A \rangle = | \Psi^A \rangle \lambda[\rho_B], \qquad (3.34)$$

where H^A is the Hamiltonian of the isolated subsystem A, $\hat{\rho}_A(\mathbf{r})$ is its density operator [the orbital representation of this operator has been used in Eqn (3.33)], and the effective potential

$$v_{\text{emb}}^A([\rho_A, \rho_B]; \mathbf{r}) = v_B(\mathbf{r}) + \int \frac{\rho_B(\mathbf{r}')}{|\mathbf{r} - \mathbf{r}'|} \mathrm{d}\mathbf{r}' + v_T^A([\rho_A + \rho_B, \rho_A]; \mathbf{r})$$

$$+ \frac{\delta E_{\text{xc}}[\rho]}{\delta\rho(\mathbf{r})}\Big|_{\rho=\rho_A+\rho_B} - \frac{\delta E_{\text{xc}}[\rho_A]}{\delta\rho_A(\mathbf{r})}$$

$$(3.35)$$

describes the effect of the environment (with potential $v_B(\mathbf{r})$ due to its nuclear contribution) on the embedded subsystem. Note that Eqn (3.34) can be used to describe both the ground state and the excited state of the embedded subsystem [for given $\{\varphi_a^A\}$ and $\rho_B(\mathbf{r})$] but, since the potential Eqn (3.35) is state-specific, the eigenvalue problem Eqn (3.34) should be solved separately for each state under consideration. Only in the case when $\rho_A^{\Psi_A}$ can be expected to be almost the same for a group of states, can all such states be determined as eigenvectors of Eqn (3.34) at once [34].

As was the case in the DFT-in-DFT scheme, the basic WFT-in-DFT Eqn (3.34) is based on the assumption that representation Eqn (3.32) for the total density is exact. This representation, however, is also correct only if the embedded orbitals $\{\varphi_a^A\}$ are orthogonal to the KS orbitals $\{\varphi_b^B\}$ of the environment. Since, only when the orbitals are orthogonal, is the absence of the crossing terms $\varphi_a^A(\mathbf{r})\varphi_b^B(\mathbf{r})$ in Eqn (3.32) justified. In order to ensure such orthogonality in practice, one again needs to expand such orbitals $\{\varphi_a^A\}$ over the set of the embedded subsystem atomic functions plus atomic functions (at least some of the functions) centered on a few nuclei of the environment important for describing the overlap region. In most wave function-based methods, the *entire* set of embedded orbitals is initially determined at the Hartree–Fock (HF) level and, then, this set is either used directly for constructing a wave function (as, e.g. in coupled cluster methods) or additionally reoptimized at the MCSCF level (as, e.g. in the MRCI or MRPT methods). In the latter case, the final MCSCF orbitals are orthonormal linear combinations of the initial HF orbitals and, hence, will automatically be orthogonal to the KS orbitals of the environment if the initial HF orbitals satisfy this condition. In analogy with Eqn (3.20a), embedded HF orbitals $\{\varphi_a^A\}$ that are required to be orthogonal to the embedded KS orbitals $\overline{\varphi}^B$ from Eqn (3.20b), can be determined from the equation

$$(1 - P^B)[F^A + v_{\text{emb}}^A([\rho_A, \rho_B])](1 - P^B)|\varphi_a^A\rangle = \eta_a^A|\varphi_a^A\rangle, \qquad (3.36)$$

where F^A is the conventional Fock operator for the isolated subsystem A and $P^B = |\overline{\varphi}^B\rangle\langle\overline{\varphi}^B|$ is the projector on $\overline{\varphi}^B$. Note that, in contrast to Eqn (3.34), the effective potential $v_{\text{emb}}^A([\rho_A, \rho_B]; \mathbf{r})$ in Eqn (3.36) involves the density $\rho_A(\mathbf{r})$, which is calculated from the diagonal quadratic form Eqn (3.10) using the N_A-lowest HF orbitals φ_a^A.

If the entire set of the embedded HF orbitals $\{\varphi_a^A\}$ from Eqn (3.36) has been obtained, then the re-optimization of these orbitals at the MCSCF level, which is required in the case of using MRCI or MRPT methods, can be performed using almost traditional generalized Brillouin conditions. The only difference is that the one-electron part (h^A) of the Hamiltonian H^A in these conditions will be replaced by the operator $h^A + v_{\text{emb}}^A$, where the potential v_{emb}^A is to be calculated using either the density $\rho_A(\mathbf{r})$ from Eqn (3.33), involving the MCSCF wave function of the lowest state or the averaged density for a group of MCSCF states. It should be emphasized that no additional requirements of external orthogonality need to be imposed on the embedded MCSCF orbitals in the course of their optimization if such requirements have already been taken into account in construction of the entire set of initial embedded orbitals, e.g. at the HF level.

3. CONCLUSIONS

The performed analysis of the requirements on orbitals of subsystems in both the DFT-in-DFT and the WFT-in-DFT approaches show that the electron density of the whole system is representable as the sum of the densities of subsystems if and, for practical purposes, only if orbitals of different subsystems are orthogonal. As a consequence, the conventional KSCED equations, which neglect explicit consideration of orthogonality requirements between subsystems, are correct only when the region where the overlap of the subsystems' orbitals is negligible (strictly speaking, only in the limit of infinitely separated subsystems). It stands to reason that in addition to errors due to an inaccurate kinetic energy functional, KSCED calculations can be expected to break down if subsystems have strong interactions such as those with non-negligible covalent character. Equations for the KS orbitals of subsystems that respect their external orthogonality were obtained. It was found that within the DFT-in-DFT scheme, knowledge of the self-consistent KS orbitals of subsystems is formally enough to reproduce completely both the exact KS orbitals for the whole system and their energies. The last statement is based on Eqn (3.23) for the matrix representation of the KS operator over the occupied KS orbitals of the entire system, which is exact for the exact $T_s[\rho]$, and, hence, this equation can be used in practice as a simple and effective test of the accuracy of approximate kinetic energy functionals.

It was also shown that in describing the embedded subsystem within an *ab initio* wave function-based method, it is enough to ensure the required orthogonality of the entire set of embedded orbitals to the KS orbitals of the environment only for the initial set of orbitals, e.g. at the HF level. Further optimization of the embedded orbitals, such as in an MCSCF procedure, will maintain orthogonality provided that unitary transformations are used for the orbital rotations.

One of the most interesting outcomes of the performed analysis is that a simple procedure was found that can be applied to KSCED procedures, including KSCED(s) and KSCED(m), to partially correct for the lack of orthogonality between subsystems. This analysis also shows that a potentially exact formulation could, in principle, be obtained with nonorthogonal orbitals, although the derivation and working equations are sure to be somewhat unwieldy. A related methodology that uses auxiliary basis sets on each subsystem to artificially enforce orthogonality can be envisaged. Further analysis may show connections between other types of corrections for nonorthogonality of subsystems, such as with cross-subsystem kinetic energy terms. One variant could involve a minimization of the average kinetic energy and would involve auxiliary basis sets (or numerical grids) with uncertain characteristics. In contrast, the analysis presented herein demonstrates a clear connection to the original KS equations and a simple, clean procedure for addressing subsystem orthogonality, either exactly in newly presented equations or approximately within broadly used equations.

ACKNOWLEDGMENTS

We gratefully acknowledge the financial support of the National Science Foundation (grant no. EPS-0814442) for the work presented herein.

REFERENCES

[1] Kohn, W.; Sham, L. J. Self-Consistent Equations Including Exchange and Correlation Effects; *Phys. Rev. A* **1965**, *140*, 1133–1138.
[2] Senatore, G.; Subbaswamy, K. R. Density Dependence of the Dielectric Constant of Rare-Gas Crystals; *Phys. Rev. B* **1986**, *34*, 5754–5757.
[3] Cortona, P. Self-Consistently Determined Properties of Solids Without Band-Structure Calculations; *Phys. Rev. B* **1991**, *44*, 8454–8458.
[4] Cortona, P. Direct Determination of Self-Consistent Total Energies and Charge Densities of Solids: A Study of the Cohesive Properties of the Alkali Halides; *Phys. Rev. B* **1992**, *46*, 2008–2014.
[5] Wesołowski, T. A.; Warshel, A. Frozen Density Functional Approach for *Ab initio* Calculations of Solvated Molecules; *J. Phys. Chem.* **1993**, *97*, 8050–8053.
[6] Wesołowski, T. A.; Weber, J. Kohn–Sham Equations with Constrained Electron Density: An Iterative Evaluation of the Ground-State Electron Density of Interacting Molecules; *Chem. Phys. Lett.* **1996**, *248*, 71–76.
[7] Wesołowski, T. A. In *Computational Chemistry: Reviews of Current Trends;* Leszczynski, J., Ed.; World Scientific: Singapore, pp 1–82.
[8] Perdew, J. P.; Levy, M. Extrema of the Density Functional for the Energy: Excited States from the Ground-State Theory; *Phys. Rev. B* **1985**, *31*, 6264–6272.
[9] Govind, N.; Wang, Y. A.; da Silva, A. J. R.; Carter, E. A. Accurate *ab initio* Energetics of Extended Subsystems via Explicit Correlation Embedded in a Density Functional Environment; *Chem. Phys. Lett.* **1998**, *295*, 129–134.
[10] Govind, N.; Wang, Y. A.; Carter, E. A. Electronic-Structure Calculations by First-Principles Density-Based Embedding of Explicitly Correlated Systems; *J. Chem. Phys.* **1999**, *110*, 7677–7688.
[11] Klüner, T.; Govind, N.; Wang, Y. A.; Carter, E. A. Prediction of Electronic Excited States of Adsorbates on Metal Surfaces from First Principles; *Phys. Rev. Lett.* **2001**, *86*, 5954–5957.

[12] Huang, P.; Carter, E. A. Local Electronic Structure Around a Single Kondo Impurity; *Nano Lett.* **2006**, *6*, 1146–1150.

[13] Liao, P.; Carter, E. A. Optical Excitations in Hematite (α-Fe_2O_3) via Embedded Cluster Models: A CASPT2 Study; *J. Phys. Chem.* C **2011**, *115*, 20795–20805.

[14] Wesołowski, T. A.; Warshel, A. Ab initio Free Energy Calculations of Solvation Free Energy Using the Frozen Density Functional Approach; *J. Phys. Chem.* **1994**, *98*, 5183–5187.

[15] Kaminski, J. W.; Gusarov, S.; Kovalenko, A.; Wesołowski, T. A. Modeling Solvatochromic Shifts Using the Orbital-Free Embedding Potential at Statistically Mechanically Averaged Solvent Density; *J. Phys. Chem.* A **2010**, *114*, 6082–6096.

[16] Neugebauer, J.; Louwerse, M. J.; Baerends, E. J.; Wesołowski, T. A. The Merits of the Frozen-Density Embedding Scheme to Model Solvatochromic Shifts; *J. Chem. Phys.* **2005**, *122* 094115/1-13.

[17] Neugebauer, J.; Jacob, Ch. R.; Wesołowski, T. A.; Baerends, E. J. An Explicit Quantum Chemical Method for Modeling Large Solvation Shells Applied to Aminocoumarin C151; *J. Phys. Chem.* A **2005**, *109*, 7805–7814.

[18] Jacob, Ch. R.; Neugebauer, J.; Jensen, L.; Visscher, L. Comparison of Frozen-Density Embedding and Discrete Reaction Field Solvent Models For Molecular Properties; *Phys. Chem. Chem. Phys.* **2006**, *8*, 2349–2359.

[19] Neugebauer, J.; Louwerse, M. J.; Belanzoni, P.; Wesolowski, T. A.; Baerends, E. J. Modeling Solvent Effects on Electron-Spin-Resonance Hyperfine Couplings by Frozen-Density Embedding; *J. Chem. Phys.* **2005**, *123* 114101/1-11.

[20] Jacob, Ch. R.; Visscher, L. Calculation of Nuclear Magnetic Resonance Shieldings Using Frozen-Density Embedding; *J. Chem. Phys.* **2006**, *125*, 194104.

[21] Pavanello, M.; Neugebauer, J. Modeling Charge Transfer Reactions with the Frozen Density Embedding Formalism; *J. Chem. Phys.* **2011**, *135* 234103/1-13.

[22] Wesołowski, T. A.; Morgantini, P.-Y.; Weber, J. Intermolecular Interaction Energies from the Total Energy Bifunctional: A Case Study of Carbazole Complexes; *J. Chem. Phys.* **2002**, *116*, 6411–6421.

[23] Jacob, Ch. R.; Wesołowski, T. A.; Visscher, L. Orbital-Free Embedding Applied to the Calculation of Induced Dipole Moments in CO_2...X (X=He, Ne, Ar, Kr, Xe, Hg) van der Waals Complexes; *J. Chem. Phys.* **2005**, *123* 174104/1-11.

[24] Kiewisch, K.; Eickerling, G.; Reiher, M.; Neugebauer, J. Topological Analysis of Electron Densities from Kohn–Sham and Subsystem Density Functional Theory; *J. Chem. Phys.* **2008**, *128* 044114/1-15.

[25] Wesolowski, T. A.; Ellinger, Y.; Weber, J. Density Functional Theory With an Approximate Kinetic Energy Functional Applied to Study Structure and Stability of Weak van der Waals Complexes; *J. Chem. Phys.* **1998**, *108*, 6078–6083.

[26] Fux, S.; Kiewisch, K.; Jacob, Ch. R.; Neugebauer, J.; Reiher, M. Analysis of Electron Density Distributions from Subsystem Density Functional Theory Applied to Coordination Bonds; *Chem. Phys. Lett.* **2008**, *461*, 353–359.

[27] Beyhan, S. M.; Götz, A. W.; Jacob, Ch. R.; Visscher, L. The Weak Covalent Bond in NgAuF (Ng=Ar, Kr, Xe): A Challenge for Subsystem Density Functional Theory; *J. Chem. Phys.* **2010**, *132* 044114/1-9.

[28] Fux, S.; Jacob, Ch. R.; Neugebauer, J.; Visscher, L.; Reiher, M. Accurate Frozen-Density Embedding Potentials As a First Step Towards a Subsystem Desription of Covalent Bonds; *J. Chem. Phys.* **2010**, *132* 164101/1-17.

[29] Dułak, M.; Kamiński, J. W.; Wesołowski, T. A. Equilibrium Geometries Of Non-covalently Bound Intermolecular Complexes Derived from Subsystem Formulation Of Density Functional Theory; *J. Chem. Theory Comput.* **2007**, *3*, 735–745.

[30] Goodpaster, J. D.; Barnes, T. A.; Miller, T. F., III Embedding Density Functional Theory for Covalently Bonded and Strongly Interacting Subsystems; *J. Chem. Phys.* **2011**, *134* 164108/1–9.

[31] Fradelos, G.; Lutz, J. J.; Wesołowski, T. A.; Piecuch, P.; Włoch, M. Embedding vs Supermolecular Strategies in Evaluating the Hydrogen-Bonding-Induced Shifts of Excitation Energies; *J. Chem. Theory Comput.* **2011**, 7, 1647–1666.

[32] Lee, H.; Lee, C.; Parr, R. G. Conjoint Gradient Correction to the Hartree-Fock Kinetic- and Exchange-Energy Density Functionals; *Phys. Rev. A* **1991**, *44*, 768–771.

[33] Parr, R. G.; Yang, W. *Density Functional Theory of Atoms and Molecules;* Oxford University Press: New York. **1989**.

[34] Khait, Y. G.; Hoffmann, M. R. Embedding Theory for Excited States; *J. Chem. Phys.* **2010**, *133* 044107/1–6.

[35] Kolos, W.; Radzio, E. Application of the Statistical Method in the Theory of Intermolecular Interactions; *Int. J. Quantum Chem.* **1978**, *13*, 627–634.

[36] Nalewajski, R. F. Integral Constraint on the Density Functional for Nonadditive Kinetic Energy in Kohn-Sham Theory for Subsystems; *Int. J. Quantum Chem.* **2000**, *76*, 252–258.

[37] Savin, A. In *Recent Developments and Applications of Modern Density Functional Theory;* Seminario, J. M., Ed.; Elsevier: London, 1996; pp 1–31.

[38] Dreizler, D. M.; Gross, E. K. *Density Functional Theory;* Springer: Berlin, **1990**.

[39] Huang, P.; Carter, E. A. Advances in Correlated Electronic Structure Methods for Solids, Surfaces, And Nanostructures; *Annu. Rev. Phys. Chem.* **2008**, *59*, 261–290.

SECTION B

Biological Modeling

Section Editor: Nathan Baker

Pacific Northwest National Laboratory, Richland, WA, USA

> CHAPTER FOUR

Structural Models and Molecular Thermodynamics of Hydration of Ions and Small Molecules

David M. Rogers*, Dian Jiao*, Lawrence R. Pratt[†] and
Susan B. Rempe*, [1]
*Center for Biological and Material Sciences, Sandia National Laboratories, Albuquerque,
New Mexico, USA
[†]Department of Chemical & Biomolecular Engineering, Tulane University, New Orleans,
Louisiana, USA
[1]Corresponding author: E-mail: slrempe@sandia.gov

Contents

Annual Reports in Computational Chemistry, Volume 8
ISSN 1574-1400,
http://dx.doi.org/10.1016/B978-0-444-59440-2.00004-1

71

Abstract

Solution equilibria are at the core of solvent-catalyzed reactions, solute separations, drug delivery, vapor partitioning and interfacial phenomena. Molecular simulation using thermodynamic integration or perturbation theory allows the calculation of these equilibria from parameterized force field models; however, the statistical many-body nature of solution environments inevitably complicates molecular interpretations of these phenomena. If our goal is molecular understanding in addition to prediction, then the statistical thermodynamic theories designed for mechanistic insight from structural analyses are especially important. In this report, we survey recent advances in the thermodynamic analysis of rigorous local structural models based on chemical structure.

1. INTRODUCTION

The possibility for solvation models based on chemical structure is founded on recent developments of a molecular quasi–chemical theory (QCT). Reviews of QCT with different technical emphases are available [1–5], and our presentation here will frequently rely on those preceding discussions. It has been explicitly argued that QCT "… transparently expresses all of the standard physical concepts that are typically used to rationalize exhaustive numerical results on the molecular thermodynamics of liquid solutions, …" [6].

Here, we focus on QCT applied to understanding the molecular thermodynamics of ions and small molecules dissolved in liquid water and protein-binding sites. As described below, these solvation properties are important to biological functions including enzymatic catalysis, signal transduction, and stabilization of molecular structure. In particular, the problem of catalytic conversion of carbon dioxide (CO_2) between gas and water-soluble forms by the enzyme carbonic anhydrase sets the stage for discussions of hydration using QCT.

1.1. Catalytic Activity of Carbonic Anhydrase

Carbonic anhydrase facilitates uptake of carbon dioxide by catalyzing the hydration of CO_2 dissolved in blood to the more soluble bicarbonate (HCO_3^-). A zinc ion (Zn^{2+}), coordinated by three histidine ligands and a water (H_2O), centers the catalytically active site (Fig. 4.1). Carbonic anhydrase is one of the fastest enzymes found in nature, with reaction rates on the order of 10^6/s [7,8]. Furthermore, the enzyme acts reversibly, catalyzing CO_2 uptake and release depending on the availability of protons

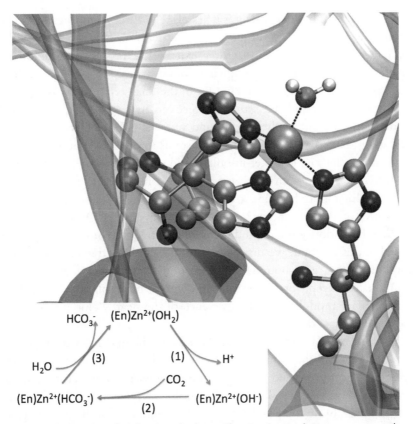

Figure 4.1 Active site of carbonic anhydrase. The zinc ion is shown as a gray sphere surrounded by the coordinating histidines and water ligands. The overall catalytic mechanism with zinc as the metal ion in the enzyme (En) active site is depicted underneath. Deprotonation of the zinc-bound water in Step 1 is thought to be rate limiting. For color version of this figure, the reader is referred to the online version of this book.

(H^+) in solution. Therefore, understanding the variations in enzyme kinetics with changes in active site chemistry, and availability of substrates (H_2O, CO_2) and products (H^+, HCO_3^-), helps clarify the carbonic anhydrase mechanism. Furthermore, insights into enzyme function could lead to efficient processes for carbon sequestration and release by robust synthetic materials. How might we mutate this enzyme to improve the catalytic properties of the zinc ion and neighboring histidines? We motivate our discussion of hydration theory with an interest in optimizing the catalytic activity of the carbonic anhydrase under conditions of varying solution pH and substrate availability.

In carbonic anhydrase, catalysis takes place in three steps (Fig. 4.1). The first step, deprotonation of the zinc-bound water molecule, is believed to be the rate-limiting step [7,9,10]. After exit of the proton to solution by Grotthus shuttling, the remaining zinc-bound hydroxide executes nucleophilic attack on a CO_2 to form HCO_3^-. In the final step, HCO_3^- is replaced by a nearby water and the active site recovers its original form. The acid dissociation constant of the zinc-bound water is 6.8 in units of pK_a [11], indicating that the water tends to lose its proton at neutral pH. In comparison, the pK_a of water in bulk liquid solution is 15.7 [12], meaning water in the carbonic anhydrase active site is far more acidic, favoring *uptake* of CO_2. Mutation of histidine to aspartate in the zinc-binding site, however, raises pK_a of this water by 2–3 units, thus favoring CO_2 *release* by the enzyme. How do we understand these shifts in pK_a? To build a computable model of this reaction, we first need to understand the free energies of the solvated proton, zinc-bound hydroxide and water, carbon dioxide, and bicarbonate. Except for the proton dissolved in bulk solution, all of these molecules are modulated by the presence of the nearby zinc metal and its coordinating ligands.

Molecular models available to analyze this problem span a vast range, and all models have distinct limitations. Born models directly coarse-grain some hydration coordinates and thereby acquire empirical parameters that necessitate validation for this application. Available *ab initio* molecular dynamics (AIMD) tools bring the possibility of higher resolution of the basic interactions. But by increasing the computational cost, AIMD calculations typically compromise the length and timescales that may be investigated. Force field models for classical molecular mechanics and simulation calculation carry traditional assumptions about the analytical form of the interactions, rigidity of some coordinates, and classical character of the atomic motions.

In the case of carbonic anhydrase specifically, a force field model for the active site zinc, which is 4-coordinate, must differ from a model of aqueous zinc, which is 6-coordinate [4,13–15]. In addition, parameterization of custom force fields for bicarbonate ion must be carried out to match the solvation properties with the particular water model used.

Transparent molecular thermodynamic assessment of these species and solution structures would be a big advantage in validating the necessary molecular modeling. That is the goal of the local structural (QCT) approach discussed here. The foundation of the QCT approach is detailed treatment of well-defined inner-shell structures. The distant, outer-shell solvent environment then becomes accessible to appropriate statistical theory.

Because the inner-shell structures are sufficiently small, electronic structure methods of computational chemistry become practical. For problems of reactivity and stability of metals in solution and in proteins, that can be an essential factor. As we will discuss below, QCT is presently the *only* available theory that can start with fully developed electronic structure calculations and proceeds through systematic statistical mechanical arguments to measurable thermodynamic properties.

1.2. Solvation Free Energies

The review below will concentrate on the evaluation of the chemical potential of Gibbs, which can be specified by $\mu_X = (\partial G / \partial n_X)_{T,p,n_{\gamma \neq X}}$. Here, T is the temperature, p the pressure, and n_γ the number (or moles) of species γ. The usual thermodynamic considerations show that the Gibbs free energy is composed as

$$G = \sum_\gamma n_\gamma \mu_\gamma. \tag{4.1}$$

From μ_X, we will extract the part that depends directly on *intermolecular* interactions, the part that vanishes when the *intermolecular* interactions vanish:

$$\beta \mu_X^{(ex)} = -\ln \langle\langle e^{-\beta \Delta U_X} \rangle\rangle_0. \tag{4.2}$$

This is the potential distribution function, which has been extensively discussed elsewhere [3]. We will confine ourselves here to only a few defining and orienting comments. As usual $(\beta k_B)^{-1} = T$, and ΔU_X is the binding energy of a distinguished molecule of type X, that is, the change of the system potential energy upon addition of that molecule. Thus, the brackets of Eqn (4.2) enclose a Boltzmann factor for the binding energy, with temperature in the familiar position. The brackets $\langle\langle...\rangle\rangle_0$ indicate the average (expectation) of the enclosed quantity over the thermal motion of the solution and the distinguished molecule when there is no interaction between these subsystems. That uncoupled status for the averaging is indicated by the subscript zero. A useful alternative formula is

$$\beta \mu_X^{(ex)} = \ln \langle e^{\beta \Delta U_X} \rangle. \tag{4.3}$$

Here, the brackets indicate a usual thermal average with interactions between solute and solution fully in play.

Further details will come out of the discussions that follow. A few broad comments are helpful here. First, we focus on the μ_X's because they, or their differences, are measurable and are direct assessments of intermolecular interactions.

Another broad comment is that the expression Eqn (4.2) is a preferred viewpoint on $\mu_X^{(ex)}$ for several reasons. First, a basic thermodynamic introduction of μ_X's often discusses specifics of thermodynamic processes that are irrelevant for the present interests. For example, we could equally well have introduced $\mu_X = (\partial A / \partial n_X)_{T,V,n_{\gamma \neq X}}$, where A is the Helmholtz free energy. These distinctions are basic to classical thermodynamics but a sidetrack for our present discussion. Simulation experts often do concern themselves with such process distinctions, but we avoid those detailed discussions here. Indeed, Eqn (4.2) or Eqn (4.3) suggests that μ_X's might be evaluated on the basis of information on the solution in the neighborhood of a distinguished solute.

A second reason that Eqn (4.2) is a preferred viewpoint on $\mu_X^{(ex)}$ is that the formula Eqn (4.2) is analogous to the partition function formulae that are familiar from introductory classes in statistical thermodynamics. That partition function structure brings advantages that will be used in the developments that follow.

1.3. Scope of This Review

The potential distribution theorem (PDT) expression [Eqn (4.2)] can lead to concise expressions for $\mu_X^{(ex)}$ on the basis of well-established thermophysical models such as the Debye–Hückel and van der Waals models. What is often missing from such models is assessment of chemical interactions at close range to a solute [16]. This issue is an important target for the QCT approaches reviewed below.

One important consequence of the lack of chemical detail in continuum models is the nontransferability of atomic Born radii (so-called) involving disparate solvents [17]. Ref. [18] argues cogently for chemically detailed treatment of dispersion interactions for different solvents and ions. The consequences of these variable dispersion interactions increase with the solute concentration. Chemical detail is even more important for di- and trivalent ions, which typically can lead to chemical distortions of ligands and nearby solvent molecules. Molecular correlations can arise from ion pairing [19] or more generic effects of short-ranged interactions. Ionic effects such as Manning counterion condensation, which plays a role in DNA packing [20,21], likely require specific consideration of ionic correlations. In Section 4, we discuss anticipated developments of Poisson–Boltzmann (PB) models from the context of the PDT to address those issues.

At the other end of the spectrum from low-resolution continuum models is direct calculation using molecular models. The potential energy surface for small molecules and molecular clusters in gas phase is routinely calculated by electronic structure methods. Comparison of calculated rotational–vibrational spectra with high-resolution spectroscopic data validates the potentials [22–24]. Because of the practical difficulty of directly solving the electronic Schrödinger equation for disordered macroscopic systems, however, significant approximations are typically required in the timescales and lengthscales of computations, as well as approximations of the electronic structure method. Characterization of the effects of those accumulated approximations is nontrivial.

Neglect of zero-point motion by approximating the atomic dynamics as classical motion on the Born–Oppenheimer energy surface also implies quantitative errors. A useful magnitude is indicated by the approximate enthalpy of transfer of Ba^{2+} from H_2O to D_2O (1.4 kcal/mol [17]). Since Ba^{2+} is a relatively heavy ion, zero-point motions of the H/D atoms are the principal concern.

Viewed in chemical detail, energies required for molecular theory and simulation are often cooperative to an interesting degree [25]. For example, using high-level *ab initio* methods on solute–water clusters, the coordinating water molecules display a nonmonotonic variation in dipole moment as the environment is enlarged [26,27] due to the nonlocal nature of the polarization. In condensed-phase applications, density functional theory (DFT) requires the addition of dispersion effects to account properly for medium- and long-distance interactions. Without dispersion corrections, solvent densities may be too low (by almost 20% for water) [28]. Effects of zero-point motions are sometimes accounted for in these simulations by an artificial increase in the system temperature [29]. This also improves the water diffusion constant, which can be 10 times smaller than its experimental value in DFT simulations [30]. These issues are complicated further where proximal phase changes are of interest, as they typically are in aqueous systems.

Advanced force field models incorporating atomic polarizabilities can be parameterized to reproduce quantum mechanical energetics. The induced–dipole interactions partially capture the effect of nonlocal dispersion [31,32]. Empirical fitting in conjunction with path-integral simulations [33–35] can approximate the effects of zero-point motion in a limited temperature range. Difficulties still remain in reproducing both gas- and condensed-phase behavior simultaneously in force field models [25]. Again, specific examples are helpful. The crystallization of classically modeled 1:1

electrolytes below 1 M concentrations [36] and the overestimation of atomic induced dipoles seen in aqueous solution force field models [37] sustain serious questions about the transferability of atomic radii and polarizabilities between different solutions.

In view of the difficulties with either molecular or continuum methods, and the practical limitations of accurate quantum mechanical models, hybrid solvation models offer advantages not available otherwise. Hybrid methods can take the form of mixed Hamiltonian models that represent short-ranged interactions explicitly and long-ranged interactions at lower levels of theory. Examples include coupling of explicit solvent models, as in quantum mechanics with molecular mechanics (QM/MM), and coupling of explicit with polarizable continuum models (QM/continuum). Reviews of progress in this field have been discussed previously in this series [38].

In the coupled Hamiltonian QM/MM models, the boundaries between regions must be closely controlled to reduce errors. QCT provides an alternative perspective to this approach by treating the boundaries as features of a statistical formulation. This statistical treatment, inherent when averaging over surrounding solvent degrees of freedom, becomes the basis for a process-oriented model of solvation described by the thermodynamic equations presented in Section 3.

2. ENERGY FUNCTION APPROXIMATIONS

Modeling of solvation requires satisfactorily accurate description of the solute–solvent interactions. Considerable effort has gone into the development of approximations with complexity increasing from point-charge and polarizable models to charge transfer and *ab initio* method development. How much complexity is required, and what is the criteria for finishing this search? In this section, we briefly discuss the issues with each method that motivate progress to more demanding approximations. We assume that the solvent model has been previously parameterized to reproduce the bulk solvent structure, density, and dielectric constant.

2.1. Lennard-Jones Function: Short-Ranged Repulsion and Longer-Ranged Attraction

The standard Lennard–Jones (LJ) form for ion-solvent interactions contains only two adjustable parameters, the coefficients c_6 and c_{12} associated with the $1/r^6$ dispersive attractive interactions, and $1/r^{12}$ overlap repulsive interactions. These terms are usually added to the Coulomb energy characterized by the ion's formal charge, q, to create a pairwise interaction of the form,

$$U_{ij}(r) = \frac{c_{12}}{r^{12}} - \frac{c_6}{r^6} + \frac{q_i q_j}{4\pi\varepsilon_0 r} = 4\varepsilon\left[\left(\frac{R_0}{r}\right)^{12} - 2\left(\frac{R_0}{r}\right)^6\right] + \frac{q_i q_j}{4\pi\varepsilon_0 r}, \quad (4.4)$$

with a radius, R_0, and an energy parameter, ε. As usual, ε_0 is the permittivity of free space involved in the Coulomb interaction. The choice of radius, R_0, is more subtle than reproducing the first peak of the ion–water radial distribution [39] since the ion–water distance can depend on the counterion [40]. Typically, strong Coulombic interactions result in average first-shell water–ion distances far into the repulsive wall. Thus, decreases in the LJ scaling lead to more favorable solvation free energies [40,41]. With fixed R_0, the choice of ε can be used to fit the free energy of ionic solvation at 298 K. This should hold over a reasonable temperature range since the entropic contribution to ionic solvation is small (~0.05 kcal/mol K [17]).

As with the Born model [3,17], Lennard-Jones radii can be solvent specific [40,42]. Transfer free energies of KCl and NaCl salts from water to formamide are 30 kcal/mol less favorable than experiment using the original OPLS-AA ion model [43], where solvents have been parameterized to reproduce pure liquid-state properties [44] and ions parameterized to reproduce aqueous ionic hydrated radius and solvation free energies [39]. Specific interactions between cations and the carbonyls of peptide backbones have been reparameterized several times [45–50] and still fail to account for differences between Na^+ and K^+ solvation between water and N-methyl acetamide or formamide [51]. Parameterizations based on reproducing both enthalpies and entropies of aqueous solvation arrive at reasonable solvated ionic radii, but the LJ model lacks sufficient freedom to reproduce both quantities consistently [41].

The ambiguities in determining LJ parameters have received critical review after the finding that almost all nonpolarizable models for sodium chloride in water showed some amount of ion aggregation below 2 M concentrations [40,52]. Crystallization was generally found to be more likely with high cation radius and low anion radius [40]. Aggregation was present at even lower concentrations for the GROMOS and AMBER force fields [36]. The explanation may lie with the geometric combining rules for LJ parameters in the GROMOS and Åqvist models [40] since geometric combinations are strictly less than arithmetic, and both underestimate the minimum energy distance for unlike pairs of noble gases [53]. In addition, the commonly used geometric combination for LJ well depths overestimates attraction between unlike pairs of noble gases [53].

Hints of solute aggregation had been seen earlier from observations of highly positive Kirkwood–Buff integrals at 2 M concentration in the GRO-MOS and AMBER force field models and somewhat positive integrals using the CHARMM force field model [54]. This behavior contrasts with experiment, which shows a somewhat negative Kirkwood–Buff integral at this concentration, indicating net *depletion* of ions in the neighborhood around a central ion. In addition, the experimental excess chemical potential of the ion pair in water increases with solute concentration at 2 M, in contradiction with results for CHARMM, GROMOS, and earlier AMBER force fields [54]. In conjunction with a recent method for estimating the osmotic pressure of force field models, it has been recommended that the interaction radius for aggregating ion pairs be adjusted separately to fit experimental osmotic pressure curves [55]. This procedure was required to model correctly the effects of osmotic stress on DNA in solutions [56]. There, the radii for all pairwise cation–anion interactions required increases from the default CHARMM27 [45], and more recent AMBER [40], ion force field models.

2.2. Short-Ranged Repulsive Interactions

One possible explanation for the lack of transferability in these force field models is the nonphysical form of the short-ranged repulsive force in the LJ potential. The LJ inner-wall often balances strong Coulombic attractions to determine solvent contacts. *Ab initio* partitioning of intermolecular energies among classical electrostatic, induction, exchange-repulsion, charge-transfer and dispersion contributions shows that repulsive effects come second in importance to classical electrostatics in determining intermolecular potential energy surfaces [57]. In other words, choosing uniform radii and solving the classical electrostatic problem is preferred over setting charge to zero and fitting atomic radii.

These problems extend to intermolecular interactions in complexes that involve the close proximity of bonded atoms from each molecule or in which the orientational dependence of the electrostatic interactions is weak. This is understandable since the r^{-n} repulsive form was introduced as an approximation for gases, where n was considered to be an adjustable parameter between 14 and 25 [58].

A single atomic radius also encounters difficulties describing distance variations and hydrogen bond strengths for complexes with strong Coulombic attractions. Furthermore, because the exchange repulsion energy depends on the overlap of occupied electronic orbitals between molecules, it has been argued that repulsive centers should be present on bonds with high electron densities as well [57]. Physically, the short-ranged repulsion should take on an exponential form [59]. The exp-6 potential is unsuitable, however, because straightforward addition of r^{-6} distorts the exponential form at short distances. A buffered 14–7 form can remedy this short-range issue.

2.3. Polarization and Charge-Transfer Effects

Many-body polarization effects can also substantially alter the water–ion interaction since the water bond angle increases from 104.52 to 106.06° [60] and its dipole moment increases in aqueous solution. For water clusters of increasing size, the dipole moment of water molecules in the first solvation shell of Cl^- first decreases and then increases toward bulk values [26]. An initial jump in the average water dipole on the addition of a second water is seen using *ab initio* methods, but not polarizable force fields, suggesting significant charge-transfer effects in the ion monohydrate [26]. Analysis of water polarizability using Hirschfeld partitioning [27] shows that the electronic polarizability of water near an ion decreases from a gas-phase value near 1.185 Å^3, toward a bulk value of 0.588 Å^3 as the cluster size increases. This same trend $(1.40–0.588 \text{ Å}^3)$ is expected as a water molecule encounters bulk liquid from a water/vacuum interface, as if the nearby ion had little influence [61]. Thus, inner-shell water–ion dipole and charge–transfer interactions are expected to be reduced in aqueous solution relative to vacuum clusters where the force field parameters are determined. This is supported by finding longer ion–water distances and lower interaction energies for monohydrates calculated using a LJ form that has been parameterized to fit bulk hydration structures and energies [39].

Polarization is likely to become increasingly important for more heterogeneous systems. This rationalizes the success in describing homogeneous liquid water properties using classical models with fixed point charges [62]. Because effective charges are optimized to reproduce average electronic structures in bulk environments, nonpolarizable force fields may encounter transferability issues in treating mixtures. Changes in molecular structure between gas and condensed phases are the most well-studied examples. A host of polarizable models have been developed that are able to capture gas to liquid increases in the water dipole from 1.85 to around 2.6 D

[35,43,63–67]. An important lesson from these studies is that the polariz-abilities of water molecules decrease in the aqueous solution. Using a single value to represent polarizability in aqueous phase then underestimates gas-phase cluster energies [66].

These decreases in polarizability result from Pauli exclusion effects. They are included in some force field models by a distance-dependent damping term that reduces the electric field experienced at a polarizable site due to nearby charges [68]. Discrepancies in the polarizability of bifurcated water chains using this method suggested the possibility that damping terms may be nonadditive [69]. Later work showed that better water cluster and chain polarizabilities could be reached within the pairwise-damping ansatz by improving the shape dependence of the water polarizability [68].

Recent quantum chemical studies of ion binding to clusters of water and small organic molecules highlight another aspect of complexity in describing solute–solvent interactions [25]. Electronic polarization contributes to both absolute and relative ion-binding free energies. Relative ion-binding free energies determine the selectivity of binding sites for a specific ion. The contribution from electronic polarization depends on the number of ligands, as noted above in the example of Cl^- hydration, as well as the specific binding site chemistry. For example, the induced dipole moments of formaldehyde molecules coordinating Na^+ or K^+ ions become less dependent on the ion with higher numbers of ligands. Because the polarization contributes substantially to the relative ion stability, subtle differences in ion size between Na^+ and K^+ produce many-body polari-zation effects critical to distinguishing these ions thermodynamically. Pairwise-additive force fields do not capture this multi-body trend.

Charge–transfer effects add another layer of complexity to molecular polarization. Although molecular polarizabilities generally decrease in condensed phases, intermolecular charge–transfer increases molecular dipole moments. This compensating effect is clearly visible in simulations of hydrated anions using polarizable force fields. Charge partitioning based on integrating the electronic density around each atom shows that the Cl^- ion loses $0.24e^-$ to the surrounding waters, while its dipole moment remains near an average of 0.6 D [26]. Polarizable models, however, consistently over-estimate the Cl^- dipole around 1.4–1.6 D [70–72]. The electronic structure of water may also be specially suited to promote electron delocalization through hydrogen-bonded structure [73], leading to complication in considering excess electrons near interfaces [74]. Although anions in water may represent an extreme case, charge buildup associated with phase potentials is a well-known difficulty of modeling electrode interfaces [75–77].

2.4. Long-Ranged Dispersion Interactions

In addition to difficulties in the form of short-ranged repulsion, direct polarization and charge–transfer terms in energy functions, there is also a fundamental link between nonadditive polarizability and dispersion interactions [53]. In the simplest picture, this connection arises because of induced–dipole/induced–dipole electronic fluctuations at frequencies near the typical femtosecond molecular simulation time step [31,78]. Although summation of r^{-6} gives the correct dimensional scaling of Hamaker coefficients for macromolecular interactions, these interactions are not additive except in the dilute limit, where the density is much smaller than $3/\alpha$, with α the molecular polarizability [79]. This nonadditivity gives rise to significant deviations to Hamaker coefficients from their additive values [80].

Dispersion interactions can dominate over electrostatic attraction between surfaces for electrolyte solutions above 0.1 M [18]. A full treatment of the theory of dispersion interactions is carried out through a separation of high frequency short-ranged interactions and low frequency long-ranged interactions between the Dirac and Maxwell fields [81], paralleling the Hartree-Fock plus excitation perturbation theory in standard quantum mechanics [57]. Simple expressions for the dispersion energy can be written in terms of molecular excitation energies or frequency-dependent polarizabilities [78] and are capable of accounting for Hofmeister trends in ionic activities [82].

Again, while simple pairwise additive forms can sometimes be parameterized to represent these dispersion interactions effectively [83], the parameters may not be transferrable between solutions of different densities. The inclusion of polarizability in force fields accounts for some of the induced-dipole/induced-dipole dispersion interactions; however, contributions from higher frequencies are usually neglected through self-consistent solution of $d_i = \alpha_i E_i$ for the induced dipoles, d_i at each time step. The induced dipole depends on the applied electric field, E, and molecular polarizability, α.

Simple energy function approximations are valuable for carrying out large simulations and extensive sampling. Classical electrostatics provides the principal contribution to intermolecular structures and energies in aqueous solutions, and the success of the available force fields follows from that.

At the same time, the transferability of these models encounters difficulties due to pair-specific contact distances, electrostatic polarization, molecular orbital overlaps that alter hydrogen-bonding properties and give rise to charge-transfer, and nonadditivities in polarization and dispersion interactions. These difficulties have decreasing energetic contributions. Unfortunately, they become prominent in gas-phase clusters, where they provoke increasing degrees of subtlety into the parameterization process. It is, however, comforting to know that pairwise-additive force fields can be constructed to match bulk and interfacial structures, energies, and thermodynamic properties once a transferrable, more complex, form is available [84,85].

3. STATISTICAL MECHANICS WITH LOCAL STRUCTURAL INFORMATION: QCT

An expanding list of physical and computational insights are being generated from the separation of local chemical interactions from the more distant influence of the environment. In this section, we develop the ideas of QCT by adding successive layers of complexity to a process starting conceptually from formation of a cavity at a specific point of interest in solution. Constraining the shape and ligand occupancy of the cavity and separating the solute-binding step from the surrounding environment allows us to construct free energy decompositions of the solvation process. The division becomes especially useful for *ab initio* molecular dynamics (MD) studies, where the computational complexity limits the calculations. Furthermore, this division permits separate analysis of complexes in different solvation environments and with different solute conformation. This separation can be helpful in analyzing relationships between structure and function in biomolecules, where active site architectures may be constrained by interactions with the surrounding system. In combination, these factors build a detailed picture of the solution properties using statistical mechanics with local structural information.

3.1. Simple Basis for QCT

A simple basis for QCT (Fig. 4.2) shows how it can be applied utilizing data streams from molecular simulations. It is a traditional [87], rough but sensible, view that such free energies can be understood by considering a process which, firstly, opens a cavity in the liquid and, secondly, places

$$\beta\mu_X^{(ex)} = -\ln p_X^{(0)}(n_\lambda = 0) + \ln\left\langle e^{\beta\varepsilon}|n_\lambda = 0\right\rangle + \ln p_X(n_\lambda = 0)$$

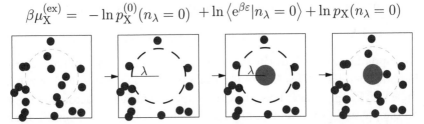

Figure 4.2 Evaluation of the excess chemical potential of the solute (red sphere) patterned according to the development of QCT. Contributions for each step (arrow) are indicated from left to right above the graphic. They are referred to as "packing," "outer shell," and "chemical" contributions, from left to right. See the text for further discussion. This formulation was written out first with the packing and outer-shell terms combined in Ref. [1], then separated as shown above in Ref. [86], and given more expansive discussion in Ref. [5]. For interpretation of the references to color in this figure legend, the reader is referred to the online version of this book.

a solute in that cavity. This is a rough view because it heavily exploits a van der Waals picture [88] of liquids, in which cavities are physically definite, and outer-shell forces are moderate so that a first-order perturbative estimate of the outer-shell contribution can be satisfactory. Our process (Fig. 4.2) goes beyond the traditional view in two crucial ways. The first way is the right-most "chemical" contribution, in which solvent molecules may refill the inner shell. Because of this, the final result does not depend on an *a priori* assignment of cavity radii. The second way our process (Fig. 4.2) goes beyond the traditional view is that the outer-shell term,

$$\left\langle e^{\beta\varepsilon}|n_\lambda = 0\right\rangle = \int e^{\beta\varepsilon}P_X(\varepsilon|n_\lambda = 0)d\varepsilon, \qquad (4.5)$$

need never be limited to a first-order perturbative (van der Waals) evaluation because simulation data will always be available to go further. Here, $P_X(\varepsilon|n_\lambda = 0)$ is the probability distribution function of the binding energy of solute X, $\varepsilon = U(\text{solvent} + X) - U(\text{solvent}) - U(X)$ in each sampled configuration.[1]

---

[1] The notation $|n_\lambda = 0$ indicates that the distribution is conditional on zero occupancy of the designated inner shell. Operationally, this means that only system configurations satisfying the condition $n_\lambda = 0$ contribute to the integral, normalized so that $\int P_X(\varepsilon| n_\lambda = 0)d\varepsilon = 1$.

In that simulation context, the generally accessible physical theory [86] is

$$k_B T \ln\langle e^{\beta\varepsilon} | n_\lambda = 0 \rangle \approx \langle \varepsilon | n_\lambda = 0 \rangle + \beta\langle \delta\varepsilon^2 | n_\lambda = 0 \rangle/2, \qquad (4.6)$$

corresponding to a Gaussian (normal) estimate of the distribution of the binding energies ε. This Eqn (4.6) should be regarded as natural first step in these QCT approaches. This estimate is expected to be satisfactory for $\lambda \sim \infty$ on the physical grounds that ε then additively combines small contributions from distant neighbors that are weakly correlated.

Dielectric – Born – models [3] of ionic hydration adopt the same form as the Gaussian model [Eqn (4.6)]. Thus, Eqn (4.6) can be regarded as a dielectric model but evaluated fully on a molecular basis. A principal limitation of conventional dielectric models is that they require parameters such as Born radii that are not typically connected to molecular character-istics by more basic theory. The statistical mechanical origin of Eqn (4.6) fixes this ambiguity of dielectric models.

The right-most, "chemical" contribution of Fig. 4.2 is so named because it can be fully analyzed on the basis of chemical concepts of serial addition of ligands to the solute. This encourages full deployment of the numerical tools of computational chemistry, and it is this that is the biggest accomplishment of QCT. We will return to discuss this issue below with the topic of ionic hydration, where chemical characteristics of ion–water interactions come to the foreground.

3.2. Polar Hydration

As a first example, we consider the hydration free energy of liquid water. Although its formulation in this context is more recent [86], we follow the path of Fig. 4.2 because of its conceptual simplicity. Water is often considered a *network liquid*, and Fig. 4.2 sheds additional light on that issue.

It is a commonplace and reasonable view that a tetrahedral H-bonding structure is the most important issue for understanding liquid water. On that basis, it is often assumed that convenient short-ranged intermolecular interactions that set those H-bonding structures are satisfactory interaction models for liquid water. We will call such models "network liquid models." Of course, general purpose interaction models for liquid water achieve tetrahedral H-bonding structures in the context of long-ranged intermo-lecular interactions; nevertheless, network liquid models are conceptually interesting.

Within the QCT formulation (Fig. 4.2), if the solute–solvent interaction energy vanishes for distances larger than λ, then the outer-shell contribution, $\ln\langle e^{\beta\varepsilon}|n_\lambda = 0\rangle$, vanishes and the full free energy can be cast as

$$\mu^{(ex)} = k_B T \ln\left[\frac{p(n_\lambda = 0)}{p^{(0)}(n_\lambda = 0)}\right]. \tag{4.7}$$

This is the *network liquid theorem:* whatever complicated solvent structure is exhibited, under the present conditions, the free energy effects are fully captured by comparison of cavity and inner-shell emptying probabilities. These contribute in opposition to each other.

Molecularly realistic simulation models of liquid water have long-ranged interactions. When the three terms of QCT (Fig. 4.2) are evaluated for such a case, the net results (Fig. 4.3) are indeed independent of the cavity radius λ employed to define this process.

Those results (Fig. 4.3) did not utilize the Gaussian approximation [Eqn (4.6)] but that physically transparent formula is surprisingly accurate (Fig. 4.4). This brings us to the observation that the ratio in Eqn (4.7) can be equal to 1 and then *packing* and *chemical* contributions balance precisely. For the case considered, this happens at $\lambda \approx 0.33$ nm. Then

$$\mu^{(ex)} \approx \langle \varepsilon|n_\lambda = 0\rangle + \beta\langle \delta\varepsilon^2|n_\lambda = 0\rangle/2. \tag{4.8}$$

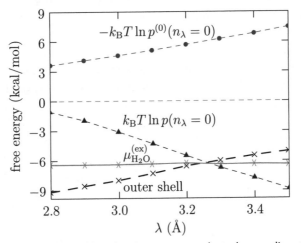

Figure 4.3 Liquid water results showing terms evaluated according to the QCT formulation of Fig. 4.2.*(Reprinted with permission from Ref. [89]. Copyright 2006, American Institute of Physics.)* For color version of this figure, the reader is referred to the online version of this book.

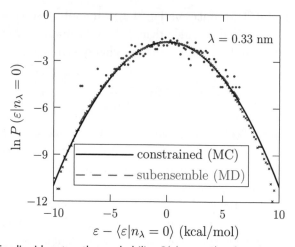

Figure 4.4 For liquid water, the probability $P(\varepsilon|n_\lambda = 0)$, when $\lambda = 0.33$ nm corresponding to the first minimum of the oxygen–oxygen radial distribution function. The smooth curves are the Gaussian models implied by the sample mean and variance. The results in red were obtained by selection of the $n_\lambda = 0$ subensemble from a standard MD calculation. *(Reprinted with permission from Ref. [86]. Copyright 2007, American Institute of Physics.)* For color version of this figure, the reader is referred to the online version of this book.

This is a surprising result because it says that a molecularly detailed dielectric model works perfectly for this exotically structured liquid if a spherical innershell with $\lambda \approx 0.33$ nm is adopted. This is to be expected if the system being considered is strongly (favorably) bound. The positive *packing* contribution dominates the result (Fig. 4.2) for small λ. Then, if the net result must be negative, we expect the negative *chemical* contribution to dominate at large λ, so the values of these two contributions should cross. For the λ at which they cross, the whole of the free energy is given by the *outer-shell* contribution.

The present formulation (Fig. 4.2) does not require inter-molecular potential models to be pair decomposable. They can be of any physical type. A landmark example of this point is that Weber and Asthagiri [90] evaluated the distribution corresponding to Fig. 4.4 on the basis of AIMD with conclusions that are consistent with the physical understanding of the Gaussian model results shown here.

3.3. Hydrophobic Hydration

The *packing* contribution to this QCT formulation (Fig. 4.2) describes the free energy cost of opening a cavity as a prospective binding site for the species of interest. If the binding interactions are not sufficiently stabilizing, then this positive *packing* contribution dominates, and such solutes in water are likely to be considered hydrophobic [91].

This *packing* contribution was first introduced for spherical cavities in the context of the scaled particle theory [92] and in the study of hydrophobic hydration [93–99]. Analysis of $\ln p^{(0)}$ ($n_\lambda = 0$) on the basis of information theory and occupancy moments directly reproduces for this contribution a partition function structure that is a signature of these quasi-chemical theories [91,100].

For a spherical solute with van der Waals interactions with an aqueous medium, the application of the QCT formulation (Fig. 4.2) works encouragingly well (Fig. 4.5) for natural choices of λ larger than the distances of closest approach of solvent neighbors. For λ smaller than that, evaluation of the *outer-shell* contribution will be difficult because intermolecular repulsive interactions with high variability will make essential contributions. Nevertheless, the results are independent of λ and do not require *a priori* assignment of cavity sizes.

It is a common rough approximation that compact solutes such as CF_4 be treated as spheres. The present formulation (Fig. 4.2) is not limited to spherical cavities, and definition of a nonspherical inner shell is necessary for careful work on such cases. It is natural to define an inner shell as a simple

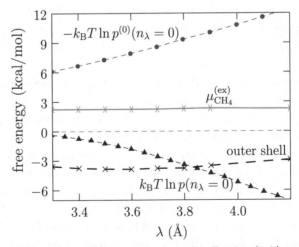

Figure 4.5 Solvation free energy contributions to CH_4. *(Reprinted with permission from Ref. [102]. Copyright 2008, American Institute of Physics. [102])* For color version of this figure, the reader is referred to the online version of this book.

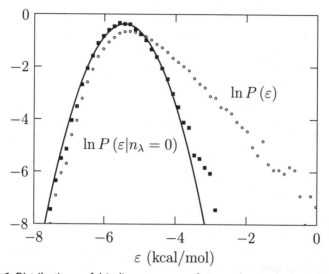

Figure 4.6 Distributions of binding energy ε for a polyatomic model of CF$_4$ in water. *(Reprinted with permission from Ref. [103]. Copyright 2007, American Chemical Society.)* Note that the conditional distribution involved in the outer-shell contribution is reasonably Gaussian, though the unconditioned distribution has a long, high-ε tail. For color version of this figure, the reader is referred to the online version of this book.

union of spherical volumes centered on the atoms. Such a strategy then eliminates intermolecular repulsive interactions from the *outer-shell* contribution. Because no *a priori* assignment of cavity sizes is required, the choice of cavity shape can be chosen to minimize the cost of directly calculating the *outer-shell* contribution. This emphasizes that the formulation Fig. 4.2 is not a van der Waals treatment, as also illustrated in the outer-shell treatment of CF$_4$ in Fig. 4.6.

It is helpful to note also that this approach can be formulated with the same physical picture utilizing a soft cutoff in defining logical quantities such as $n_\lambda = 0$ [101]. This permits all necessary calculations to be implemented with MD in addition to Monte Carlo methods.

As a solute in water, CO$_2$ is neither polar, nor hydrophobic, nor ionic. But this is a case of practical importance as our introduction indicates, and, in addition, CO$_2$ often exhibits chemical interactions with an aqueous medium. For the three-atom, explicit partial charge model of CO$_2$ [104], similar statistical thermodynamic results have been found (Fig. 4.7) with the formulation of Fig. 4.2. In this case, the mean-field nature of the solvent distribution around CO$_2$ was exploited to approximate the *outer-shell* contribution on the basis of a Born model plus dispersion interactions.

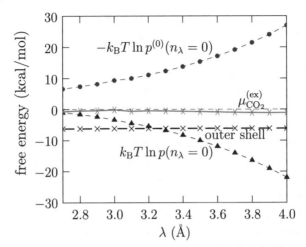

Figure 4.7 Solvation free energy contributions to CO_2. *(Reprinted with permission from Ref. [104]. Copyright 2011, American Institute of Physics.)* The total is nearly independent of the cavity radius. Note that physically distinct contributions of magnitude 10 kcal/ mol balance to produce a correctly sized result of less than 1 kcal/mol. For color version of this figure, the reader is referred to the online version of this book.

The packing and chemical terms have been estimated roughly by extrapolation of the (nonspherical) cavity occupancy distributions to $n = 0$ [93]. Even with these approximations, this estimate of the CO_2 hydration free energy (0.06 kcal/mol) shows good agreement with experimental data (0.24 kcal/mol) [105]. The total (Fig. 4.7), nearly independent of cavity size, shows that the mean-field approximations employed worked satisfactorily for this case. This approach applied to earlier work on H_2 hydration similarly showed results independent of cavity size [106,99]. We note here that zero-point motion plays a role in determining the chemical potential of H_2.

3.4. Extension: More Conditioning

The development of Fig. 4.2 generalizes naturally to conditions $n_\lambda = n > 0$. Derivations are given elsewhere [5].

Considering results for all nonnegative integers, n, leads to a partition function formula for the free energy $\mu_X^{(ex)}$ [5]. It has been specifically argued [5] that the $n_\lambda = 0$ is preferred over $n_\lambda > 0$ because (1) the packing contribution has a helpful interpretation and substantial extant theory [93–99]; (2) the chemical contribution similarly has a helpful physical interpretation; and finally (3) the outer-shell contribution is simplest; indeed, it may be usefully approximated by the Gaussian model of Eqn (4.6) because close neighbors are not involved. The latter point is not only convenient, but it also contributes to building a physical under-standing of these phenomena as Figs 4.4 and 4.6, and Eqns (7) and (8) show. Indeed the payoff for the specific applications of Fig. 4.2 is the simplicity of Figs. 4.4 and 4.6.

With the $n_\lambda = 0$ condition, the *chemical* contribution is naturally viewed in the inverse way—in terms of chemical processes that combine ligands with the solute surrounded by an evacuated inner shell. This gives immediate access to the tools of computational electronic structure calculations. This issue has been discussed in detail [5] and is followed up in the next section.

The physical reason for the insight noted above is that the formulation Fig. 4.2 focuses on the range λ of the interactions with the solute. The range λ of the interactions has long been recognized as basic to the physical understanding of solutions [107,108].

The formulation Fig. 4.2 suggests the advantage of identifying structural conditions that clarify mechanisms of interest. In the context of molecular simulation calculations, this *conditioning* idea was tried first for electrostatic contributions to the hydration free energies of atomic ions in water [109], with encouraging success. That work was a direct motivation of the modern development of molecular QCT [110,111].

It is interesting here to note more exotic conditioning concepts. One idea is to condition the binding energy distributions on a value of the extreme pair interaction, assuming a reasonable pair-decomposable inter-action model. This idea is suggested by the fact that extreme contributions seem to be the difficulty with unconditioned distribution of Fig. 4.6. Focusing on the largest (most unfavorable) contribution, $\varepsilon_{\max}$, leads to the conditional distribution $P(\varepsilon|\varepsilon_{\max})$. For liquid water that approach works moderately well [112], but not better than the direct use of Fig. 4.2, which is after all surprisingly successful.

3.5. Conditioning for Structural Analysis

A condition targeting a structured constellation of n ligands (Fig. 4.8) generalizes Fig. 4.2 and addresses ion association to protein-binding sites. This more collective condition may be denoted by C_n, an indicator function mapping the inner-shell structure to a Boolean value of true or false. $C_n = \text{TRUE}$ if C_n were enforced as a constraint. n indicates that the condition is based on an $n_\lambda = n$ ligand geometry. C_n can select for a narrow

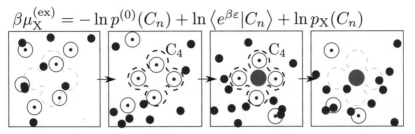

$$\beta\mu_{X}^{(ex)} = -\ln p^{(0)}(C_n) + \ln \left\langle e^{\beta\varepsilon}|C_n\right\rangle + \ln p_X(C_n)$$

Figure 4.8 The excess chemical potential of X (red sphere) described by QCT as solute binding within a preformed site. The condition C_n ensures that the n ligands (open spheres) form the specified geometry. The free energy terms refer to the probability for spontaneous *binding site formation* in the absence of the solute, the free energy for *solute binding*, and the free energy for *constraint release*. These are steps A → B → E → F of thermodynamic cycle 2 from Ref. [113]. For interpretation of the references to color in this figure legend, the reader is referred to the online version of this book.

set of structures, as shown schematically in Fig. 4.8 by small deviations of the ligand positions within the binding site. Conditions that address physical mechanisms are the target of these developments. C_n proves particularly helpful when studying molecular binding sites formed in macromolecules.

Thus, C_n is a structural condition [113]. It permits assessment of *architectural* control of a *binding site*, where interactions with the surrounding environment may constrain ligand composition and configurations. The left-most step of Fig. 4.8 identifies a free energy contribution for the surrounding matrix to arrange ligands in an anticipated structure. The middle term provides the free energy for *solute binding*,

$$X + L_n \rightleftharpoons X \cdot L_n \tag{4.9}$$

in the $C_n = \mathrm{TRUE}$ sub-ensemble and in the presence of the surrounding environment. Comparing this term for different solutes gives the *selectivity* of the C_n site after the environment has exerted its influence over initial formation of that binding site. As we discuss below, standard results of statistical thermodynamics show how to evaluate that chemical potential [3,4] and, with such a strongly specified configuration, computational electronic structure calculations can be of immediate use. The right-most step (Fig. 4.8) gives the free energy for *constraint release*, obviously associated with the free energy cost of imposing the C_n condition on the ligand structure within the solute-occupied binding site.

Refs. [102,113] used the notation $p_0(C_n)$ to indicate the probability for *binding site formation* in the absence of solute. The present notation treats flexible, i.e. molecular, solutes without further adjustments.

Thus, these conditioning ideas aim to elucidate physical mechanisms of solvation processes. C_n is a structural condition designed to quantify the role of the environment's preferences for specific binding site compositions and configurations.

An *architectural control mechanism* has been used to rationalize selective binding of K^+ over Na^+ in the ion carrier, valinomycin [119], and in potassium ion channels [51,114–116]. In valinomycin [117], a cyclic macromolecule, crystal structure data shows K^+ in binding sites coordinated by $n = 6$ carbonyl oxygen ligands, in contrast to $n = 8$ carbonyls, or a mixture of hydroxyls and carbonyls, in potassium ion channels [118]. A long-standing puzzle relevant to both molecules is how a single binding site, or sites with similar chemical motifs, can be strongly or weakly selective and what role the environment plays in the selectivity. In the valinomycin example, selectivity is sensitive to the environment, i.e. the solvent properties. Figure 4.8 applied with varied C_n provides for quantitative investigation of architectural control in these solvation problems.

To illustrate this idea, we used molecular simulation to decompose the total relative solvation free energies $\left(\Delta\mu^{(ex)} = \mu_{K^+}^{(ex)} - \mu_{Na^+}^{(ex)} \right)$ upon replacement of Na^+ with K^+ in model binding sites. The differencing between ions does not require an absolute solvation free energy, but in principle, those absolute values could be obtained also. In a relative calculation of solvation within the same binding site, the first term cancels. Thus, this difference depends only on the two right-most contributions of Fig. 4.8 (see Fig. 4.9 and Ref. [113]).

In this example, n water molecules formed the binding site. To define the binding site complex, we applied a flat-bottomed harmonic potential to the n ligand oxygens, starting at 3.1 A from the ion. We treated the outer environment as a vacuum. The vacuum serves as an important reference for defining environmental perturbations that may influence the structure of the binding site. Acknowledging the influence of the environment to be investigated, we specified ligand geometries, C_n, to be separately investigated. Here, we chose a condition that selects small deviations, quantified by root-mean-squared displacements (RMSD), from defined reference structures. Our choice of reference structures consisted of minimum energy geometries of n ligands centered on an ion. Thus, when C_n is tightly enforced, binding-site architectures *fit* a specific ion. The notation of Fig. 4.9 uses the (C_6, Na^+) and (C_6, K^+) to indicate structures centered on Na^+ (left) and K^+ (right) for $n = 6$.

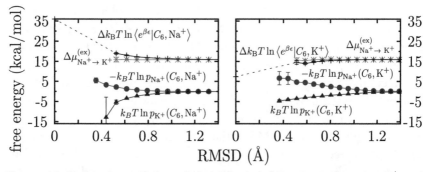

Figure 4.9 Components of the relative solvation free energy between K^+ and $Na^+ \Delta\mu^{(ex)}$ (green X) depend on confinement of binding site architecture of $n = 6$ water clusters. C_n selects for ligands taking positions closer to their minimum energy configurations near Na^+ (left, at small RMSD) and increases the cost of replacing Na^+ by the larger K^+ ion. Formation of this coordination structure with Na^+ is more probable than with K^+ (ln $p_{Na^+} >$ ln p_{K^+}). This illustrates the traditional steric mechanism for creating a binding site favoring Na^+ ($\Delta k_B T$ ln$\langle e^{\beta\varepsilon}|C_n\rangle > \Delta\mu^{(ex)}$). This trend can be reversed using an $n = 6$ coordination structure favored by K^+ (right panel). *(Adapted with permission from Ref. [113]. Copyright 2011, American Chemical Society).* For interpretation of the references to color in this figure legend, the reader is referred to the online version of this book.

The free energy decomposition (Fig. 4.8) is structural, creating three contributions for this case of differencing in ion solvation: a relative *solute binding* free energy, $\Delta\ln\langle e^{\beta\varepsilon}|C_n\rangle = \ln\langle e^{\beta\varepsilon}|C_n\rangle_{K^+} - \ln\langle e^{\beta\varepsilon}|C_n\rangle_{Na^+}$, for each chosen n, along with constraint release, $p_{K^+}(C_n)$ and $p_{Na^+}(C_n)$. We varied the allowed structural deviation (RMSD) from each reference structure, which is analogous to varying the inner-shell cavity size, λ. As in Figs 4.3, 4.5, and 4.7, the relative *solute binding* free energy and the difference in *constraint release* free energies for K^+ and Na^+ sum to a constant total $\Delta\mu^{(ex)}$ independent of RMSD (Fig. 4.9).

The free energy of ion selectivity of a C_n site vs. bulk water is given by the difference between $\Delta\ln\langle e^{\beta\varepsilon}|C_n\rangle$ in the constrained system and the relative bulk hydration free energy of the two ions. For the force field model used in this application, the $Na^+ \rightarrow K^+$ hydration free energy difference for ion substitution in bulk water is 17.3 kcal/mol. Thus, nearly all C_n complexes weakly select for K^+ in the absence of other conditions since $\Delta\mu^{(ex)} < 17.3$ kcal/mol in Fig. 4.9. What happens to selectivity of the binding site complex when the environment enforces a specific structural constraint, C_n? QCT provides a quantitative answer to this question and indeed appears to be unique in this ability.

For large values of RMSD at the right side of the plots in Fig. 4.9, the relative *solute binding* free energy $\Delta\ln\langle e^{\beta\varepsilon}|C_n\rangle$ approaches the total free energy for ion binding to the complex, $\Delta\mu^{(\text{ex})}$. In this limit, the complex is not highly selective vs bulk water, yielding observation of a variety of ligand configurations, with more diversity around the smaller Na^+. When no RMSD constraint is imposed to enforce specific ligand geometries (C_n), the constraint free energies are zero: $\ln p_{K^+} = \ln p_{Na^+} = 0$.

Upon tightening the RMSD, the total free energy of ion selectivity, $\Delta\mu^{(\text{ex})}$ (solid green line on Fig. 4.9), is split by the coordination condition, C_n. Now, the three free energy contributions are evident, and $\Delta\ln\langle e^{\beta\varepsilon}|C_n\rangle$ represents the change in *solute binding* free energy upon replacing Na^+ with K^+ *within the preformed complex*, i.e. only when C_n is TRUE.

In highly constrained states that enforce the reference ligand configuration centered around K^+, (small RMSD in the right plot of Fig. 4.9), the change in solute binding free energy upon replacing $Na^+ \rightarrow K^+$ becomes less costly and thus more favorable for K^+. See the downward trend of $\Delta\ln\langle e^{\beta\varepsilon}|C_n\rangle$ as RMSD decreases. Furthermore, Na^+ becomes less likely than K^+ to occupy these geometries as the water cluster becomes more constrained, even though Na^+ is the smaller solute. See the sharper upward deflection of the $p_{Na^+}(C_n)$ compared to the downward deflection of $p_{K^+}(C_n)$. In contrast, the same water-cluster binding site can be tuned in the other direction, to select for Na^+, by an environment that imposes the appropriate Na^+-centered structural condition. For example, the $C_6 = \text{TRUE}$ condition centered on Na^+ for small RMSD (left plot of Fig. 4.9) excludes configurations with more distant ligand positions favored by the larger K^+. This illustrates a traditional steric mechanism of selectivity.

Figure 4.9 illustrates the idea that the external environment can change the ion selectivity of a single binding site. The structural decomposition used gives different architectures with different selectivities. If the environment provides an energetic incentive for adopting a particular architecture, the selectivity can be shifted in proportion. In valinomycin, for example, a mechanism for shifting the binding site architecture is through altering the strength of intramolecular hydrogen bonds [51,119]. The free energies for *binding site formation* and *constraint release* directly quantify this incentive. The thermodynamic cycles of Ref. [113] give multiple routes for its calculation.

To summarize, the structural conditioning (Fig. 4.8) assesses *architectural control* of solute free energy. This assessment works for definitions of structural conditions enforced loosely or rigidly. Application to simple ion–water complexes highlights the basic feature: selectivity for a specific ion can be tuned by an environment that enforces constraints on binding site architectures to conformations favored by that ion but biased well away from structures favored by other ions (Fig. 4.9 and Ref. [113]). This approach (Fig. 4.8) obviously applies to binding sites in general and is currently being used to assess the role of the environment in determining solute free energy in larger models of macromolecular binding sites (Rogers and Rempe, unpublished). The next sections describe closely related formulations of QCT used in earlier work [114,119–120], in the context of K^+-selective ion channels and macromolecules, that played a key role in the original *architectural control* idea.

A final technical observation is clear from the free energy decomposition of Figs 4.8 and 4.9. The constraint release contributions, $\ln\ p_X(C_n)$–$\ln\ p_{X'}(C_n)$, are important in this balance. They combine with the solute binding contributions to form a constant total relative solvation free energy $\Delta\mu^{(ex)} = \mu_X^{(ex)} - \mu_{X'}^{(ex)}$.

This constraint release contribution has been overlooked in some recent discussions of the mechanism of ion selectivity in protein-binding sites [121a]. The result is an incomplete assessment of selectivity.

3.6. Ion Hydration and Uncoupling from the Environment

Simple models of solute binding reactions exploit the vapor-phase partition function of the isolated molecules, X and L, and the complexes they form. The structural condition, C_n, precisely defines the composition, n, and configuration of the poly-molecular complex, XL_n.

The two left-most arrows of Fig. 4.8 indicate evaluation of the interaction contribution to the free energy of formation of the species XL_n defined by C_n. Evaluation of that free energy of formation is the subject of the well-developed theory of chemical equilibrium [3,122,123].

It is always possible, and a common procedure, to separate from that chemical equilibrium evaluation the contribution obtained when there are no intermolecular interactions of the XL_n complex with an external medium. This ideal step is a computational convenience addressing the nontrivial problem of the XL_n complex alone; the temperature and densities of all species in the solution are those of actual physical interest and then the remainder—or *intermolecular interaction part*—is addressed at the last.

The central step in this procedure leads to the free energy of *solute binding* within the inner shell and in an ideal state $-k_B T\ \ln K^{(0)} \rho_L^n$, as worked out in standard texts [122] and illustrated in the second term of

$$\beta\mu_X^{(ex)} = -n\beta\mu_L^{(ex)} - \ln[K^{(0)}(C_n)\rho_L{}^n] + \beta\mu_{X\cdot L_n}^{(ex)} + \ln p_X(C_n)$$

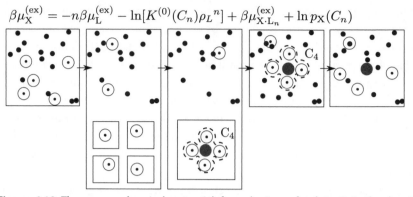

Figure 4.10 The excess chemical potential for solvation of solute X (red sphere) described by QCT as *ligand removal* from the environment, an ideal *solute binding* with the $X\cdot L_n$ uncoupled from the solution, followed by *restoration* of the $X\cdot L_n$ solute complex with the environment, and *constraint release*. The condition, C_n, ensures that the ligands, L (open spheres), make up the specified coordination. These correspond to steps A → B → D → E → F of cycle 1 in Ref. [113]. For interpretation of the references to color in this figure legend, the reader is referred to the online version of this book.

Fig. 4.10. The densities, ρ_L, are the densities of the ligands in the solution. Viewed physically, these density factors account for ligand availability in the solution. The other terms shown in that figure give the free energies for *removing ligands* from the starting phase $-\mu_L^{(ex)}$ and for *replacing the complex* into that phase $-\mu_{X\cdot L_n}^{(ex)}$. Finally, the right-most term of Fig. 4.10 is the *constraint-release* contribution, familiar already from Figs. 4.2 and 4.8. The explanation here is a paraphrase of the more detailed arguments of Refs. [1,5] articulated specifically for the n condition considered in the earlier work.

The separate consideration of the long-ranged interactions during solvent *removal* and *insertion* of the complex into the solvation environment isolates the role of solvent optical properties and chemical potential in driving the long-ranged aspect of the solvation process. These long-ranged contributions balance some of the unfavorable cavity formation free energy for hydrophobic solutes (Figs. 4.3 and 4.5). In protein-binding sites, they become especially important in controlling solute coordination preferences, as described in the preceding section (Fig. 4.9).

The chemical equilibrium constant, $K^{(0)}(C_n)$, isolates the local interactions responsible for solute activity. Once a definite C_n is adopted, evaluation of $K^{(0)}(C_n)$ is an application of the well-established theory of chemical equilibrium. This step gives rise to contributions associated with ligand field strength and is appropriately modeled using quantum mechanics [1]. Because the formula (Fig. 4.10) must produce a constant, $\mu_X^{(ex)}$, for any choice of n, the variation with n in the first three terms of Fig. 4.10 indicates the relative population, $p_X(C_n)$, of each coordination state composed of solute X and n ligands, L.

In some of the earliest applications of QCT, a series of cation–water structures were energy minimized in gas phase, providing separate estimations of $K^{(0)}(C_n)$ using the DFT formulation of QM [128,129,130,131,15]. A Poisson electrostatic model approximated the outer solvation environment for both the removal of water ligands from the environment, $-\mu_L^{(ex)}$, and the restoration of the ion–water complex, $\mu_{X \cdot L_n}^{(ex)}$ (Fig. 4.10). Because the same approximation is used for water *removal* and *restoration*, boundary problems arising from the choice of water atom radius cancel to a large degree.

Also, coordination structures were highly constrained to *simple*, single, dominant, coordination geometry (C_n) centered around each metal cation. *Simple* structures that lack *split-shell* coordination character appear for the first four (4) waters near the solute in simulations of metal ions in water (Fig. 4.11a). Note that these *simple* structures may consist of a subset of structures that define the full first peak of radial distribution functions, as for Na^+ and K^+ (Fig. 4.11a). By choosing these structures for QCT analysis (Fig. 4.10), the approximation $\ln p_X(C_n) \approx 0$ is nearly exact for the n with the highest $\ln p_X(C_n) \approx 1$, or equivalently, the lowest $-n\mu_L^{(ex)} - k_B T \ln[K^{(0)}(C_n)\rho_L^n] + \mu_{X \cdot L_n}^{(ex)}$. We denote the most probable composition by $\bar{n}$. This defines the *primitive approximation* to Fig. 4.10. As an approximation to QCT, it should not be confused with QCT itself, which is an exact statistical mechanical theory of chemical equilibrium.

Simple coordination geometries were selected by setting inner-shell boundaries, λ, according to *ab initio* molecular simulation results (Fig. 4.11a). For Li^+, Na^+, and K^+, λ was set so that $n = 4$ dominates the inner-shell structure in liquid water simulations, and checked for consistency in the QCT calculations.

Estimates of absolute solvation free energy, $\mu^{(ex)}$, from this process (Fig. 4.10) agree remarkably well with experiment over a large range of monovalent and divalent cations, including transition metals (Fig. 4.11b) [15,132]. We highlight two interesting observations based on these results, which are also relevant to proteins, including potassium ion channels. The free energy components for Li^+, Na^+, and K^+ as a function of n (Fig. 4.12) illustrate the observations.

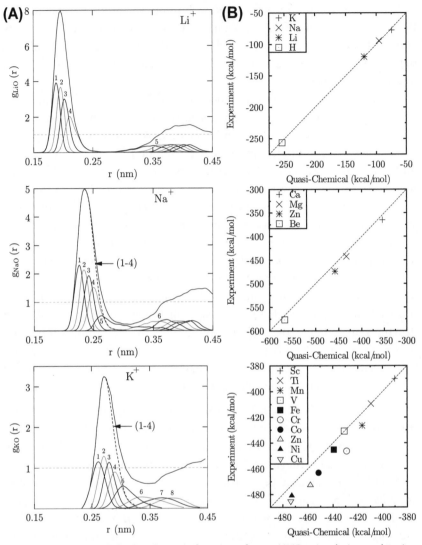

Figure 4.11 (A) Radial distribution function from AIMD simulations showing distribution of water oxygens sorted by distance from the ion. Waters that stay within the first peak define *simple* coordination complexes that lack a split-shell coordination structure, making them ideal for QCT analysis. These simple coordination structures consist of up to four (4) waters for Li^+ [124–126], Na^+, and K^+. *(Reprinted from Ref. [114], Copyright 2007 with permission from Elsevier).* (B) QCT vs experimental results for ion hydration, $\mu^{(ex)}$.Values on the diagonal indicate perfect agreement. *(Adapted with permission from Ref. [15]. Copyright 2004 American Chemical Society.)*

First, the most probable ligand composition within the inner shell, $\bar{n}$, shifts depending on the component solvation process (Fig. 4.12). *Solute binding* in an ideal vapor phase $(\ln[K^{(0)}(C_n)])$, with low ligand density of $\rho_{H_2O}k_BT = 1$ atm, yields $\bar{n} = 4$. The value of $\bar{n}$ increases for solute binding after accounting for an enhanced ligand availability typical of condensed phases $(\ln[K^{(0)}(C_n)\rho_{H_2O}{}^n]$ for $\rho_{H_2O}k_BT = 1354$ atm). In liquid phase $(\mu^{(ex)})$, the most probable composition decreases to $\bar{n} = 4$ after including contributions from the outer solvation environment $(-n\mu_L^{(ex)} + \mu_{X \cdot L_n}^{(ex)})$. This last shift occurs because long-ranged interactions from a high dielectric medium, like liquid water, favorably solvates ligands, L, and ion–ligand complexes, XL_n. The free energy penalty for removing ligands $(-n\mu_L^{(ex)})$ from this outer environment increases with n and thus drives $\bar{n}$ down to lower coordinations [114,121].

Solvation processes that stabilize high vs. low ligand compositions, $\bar{n}$, provide insights applicable to macromolecular binding sites. Although high $\bar{n}$ within the first coordination shell is rare in liquid water (Fig. 4.11a), it is common in protein-binding sites [121]. For example, crystallized potassium ion channels depict twice as many ligands centered around K^+ $(\bar{n} = 8)$ [118] as found within the same inner-shell volume in water (Fig. 4.11a). Figure 4.12 reveals a solvation process that stabilizes high $\bar{n}$: $\ln[K^{(0)}(C_n)\rho_L^n]$, that is, *solute binding* in the presence of high ligand availability coupled with an outer environment that poorly solvates the ligands $(\mu_L^{(ex)} \sim 0)$.

A bioinformatics analysis comparing structures of highly selective potassium ion channels noted low density of hydrogen-bond donating groups near the binding sites [121]. Hydrogen-bond donors would favorably compete for carbonyl ligands. The absence of hydrogen-bond donors, coupled with high ligand availability, led to the suggestion that these criteria can also stabilize high $\bar{n}$ in proteins [121]. This solvation process was referred to as "phase activation," and the physical characteristics that stabilized high coordination were called a "quasi-liquid" phase [114,121].

A second observation (Fig. 4.12) is that $\mu^{(ex)}$ in liquid water includes a prominent contribution from the outer environment $(-n\mu_L^{(ex)} + n\mu_{X \cdot L_n}^{(ex)})$, in addition to the dominant chemical contribution $(\ln[K^{(0)}(C_n)\rho_L^n])$ [128,129,131]. For ions to transfer spontaneously from water to protein-binding sites, ion solvation in the protein should be at least as favorable as ion hydration free energy, $\mu^{(ex)}$. In fact, a matching of free energies facilitate ion

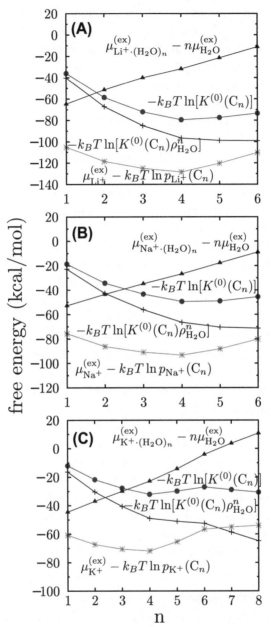

Figure 4.12 Free energy components according to Fig. 4.10 for formation of ion–water complexes in liquid water as a function of n waters around ions: (A) Li^+, (B) Na^+, and (C) K^+. Inner-shell boundaries are set according to Fig. 4.11a to yield simple $n = 4$ configurations as the dominant compositions in liquid phase. Estimating $p_X(C_n) \approx 1$ for the dominant structure provides reasonable estimates of $\mu^{(ex)}$ (see Fig. 4.11B). *[(A) adapted with permission from Ref. [128]. Copyright 2000, American Chemical Society. (B) and (C) plotted from data adapted with permission from Ref. [121]. Copyright 2008, American Chemical Society.]* For color version of this figure, the reader is referred to the online version of this book.

transport. If contributions from the outer environment are small, then increases in the number n of binding site ligands, or the strength of the ligand dipoles, may make up the free energy difference. Highly selective potassium ion channels appear to utilize both: $\bar{n}$ and carbonyl dipoles in binding sites that, relative to water, overcoordinate K^+ using ligands with stronger dipoles [114]. This contrasts with the traditional proposition that channel-binding sites conducive to transport would mimic K^+ hydration structures.

The first application of QCT (Fig. 4.10) to a model potassium ion channel-binding site (Fig. 4.13) confirmed the predictions described above: the environment plays an important role in stabilizing crowded binding sites (high $\bar{n}$) that facilitate K^+ transport. More importantly, QCT analysis (Fig. 4.10) provided among the earliest evidence [120] for *architectural control* of binding sites for Na^+ rejection and for tuning selectivity of binding sites with identical chemical motifs [114,119,121].

Diglycine (GG) molecules embedded in a low dielectric medium provided a structurally relevant model. Diglycine represents monomeric units of channel protein that assemble within lipid bilayers, in the presence of high K^+ concentration, to form a tetramer centered around K^+ with $\bar{n} = 8$ carbonyl ligands. Note that each channel monomer, and each diglycine molecule, provides two (2) carbonyl ($C=O$) ligands spaced appropriately for K^+ coordination. The energy-minimized structure of diglycine tetramer centered on K^+ closely resembles the crystal structure of a highly selective

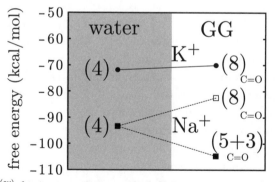

Figure 4.13 $\mu^{(ex)}$ for ion association with (left) $\bar{n} = 4$ waters in liquid water and (right) a channel binding site modeled as a tetramer of bidentate diglycine (GG) molecules in low dielectric surroundings. A diglycine binding site constrained to be centered on K^+ ($\bar{n} = 8$ carbonyls) destabilizes Na^+, whereas a site centered on Na^+ ($\bar{n} = 5+3$ carbonyls) favors Na^+. *(Adapted from Ref. [114], Copyright 2007 with permission from Elsevier.)* For color version of this figure, the reader is referred to the online version of this book.

potassium channel [118]: structural deviation from an interior K^+ channel-binding site (S2) is only 0.4 Å (RMSD).

The diglycine binding site also reproduces functional properties of potassium channels (Fig. 4.13). The absolute value for K^+ solvation $\mu_{K^+}^{(ex)}$ in diglycine tetramer centered on K^+ ($\bar{n} = 8$) resembles the predicted K^+ hydration free energy. Hence, the overcoordinated model binding site is conducive to K^+ transport [114]. The same K^+-centered ($\bar{n} = 8$) diglycine structure substantially destabilizes Na^+ compared to water. Thus, ligands constrained by the surrounding environment to this $\bar{n} = 8$ configuration favored by K^+ yield a binding site highly selective for K^+. High K^+ selectivity persists even in a binding site that shrinks closer to solvate the smaller Na^+ ion, or expands to greater distances, as long as all $n = 8$ ligands remain at about equal distances from the ion [114].

In contrast, the dominant structure for diglycine tetramer centered on Na^+ exhibits large structural deviation from K^+-occupied binding sites. In this case, a split-shell structure is favored ($\bar{n} = 5 + 3$). Furthermore, this ligand configuration favorably solvates Na^+ relative to water (Fig. 4.13). Thus, a binding site capable of providing an accessible Na^+-centered configuration can block K^+ transport by binding Na^+, even while being conducive to K^+ transport in the absence of Na^+. Note that computation of absolute $\mu^{(ex)}$ (Fig. 4.13) facilitates comparison of ion binding to sites with different ligand composition and configuration.

The *architectural control* mechanism, illustrated in Fig. 4.13 and Fig. 4.9, helps interpret a variety of observations: high K^+ selectivity associated with K^+ binding to K^+-centered crystal structures [118]; block associated with Na^+ binding to Na^+-centered structures [133,134]; binding site composition and configuration, or channel selectivity, dependent on hydrogen bond donors or acceptors in the proximal environment [135–139]; and high or low selectivity in binding sites with identical chemical motifs [140].

These earliest QCT studies of selectivity applied to potassium channels (Fig. 4.13), and later valinomycin, provided some of the earliest and still most convincing arguments for the prominent role of the environment in determining selectivity of ion-binding sites [114,119–121]. Extension (Fig. 4.10) to include multiple binding sites and multiple ion occupancy within the inner shell may provide additional mechanistic details (Varma and Rempe, unpublished). A comparison of potassium channels with valinomycin highlights two distinct modes of *architectural control* used to achieve K^+ selectivity. An advantage [114,119,121] of the highly crowded $\bar{n} = 8$

binding sites of potassium channels compared to valinomycin ($\bar{n} = 6$) is that high selectivity persists even in the fluctuating binding site due to the lack of Na^+-centered $\bar{n} = 8$ structures (Figs. 4.11a and 4.12).

A common approximation treats vibrational motions harmonically in calculations of the free energy of *solute binding* (Fig. 4.10). Contrary to some assertions [141], this is not required. The topic of anharmonic motion is well studied [24] and can be avoided for highly constrained molecular structures with small ion-ligand distances [15,114,121,129,131]. Methods that are readily available account for anharmonicity and multiple anharmonic minima for cases of weakly bound complexes and larger n [99,142,143]. A path-integral approach [144] provides a general method for treating anharmonicity in coordination complexes.

The spirit of the QCT free energy separation, however, is to identify tightly associated ligands that form simple coordination structures appropriately modeled by harmonic potentials (Fig. 4.11a) and treat the remainder using statistical mechanics. The ion hydration results (Fig. 4.11b), as well as results based on Fig. 4.10 for the more weakly bound solution structure of CO_2 [104], attest to the utility of this approach.

3.7. Ligand–Protein Association

The coordination complex picture of QCT applies equally well to solvation by biomolecules and inhomogeneous environments in general. Combining the local structural thermodynamic model with the effects of the environment surrounding macromolecule binding sites allows us to identify the influence of external energetic and structural conditions on the behavior of solute coordination [51]. In other words, we can create statistical mechanical models utilizing coordination complexes as building blocks.

It is easy to see how Figs 4.8 and 4.10 can be combined (Fig. 4.14) to study interactions with both nearby binding site ligands and solvent molecules free to move throughout a system. The cost of removing a solvent molecule W from solution contains a translational contribution to the chemical potential, μ_W. For tethered or protein-bound ligands, this translational component turns into the log–likelihood of adopting the expected conformation, $k_B T \ln p^{(0)}(C_n)$. The translational cost plus the cost of energetically decoupling the solvent/ligand molecules from solution constitute the "desolvation penalty" for moving them into the binding site. Note also

$$\beta\mu_{\mathrm{X}}^{(\mathrm{ex})} = -\ln p^{(0)}(C_n) - \beta(\mu_{\mathrm{L}_n}^{(\mathrm{ex})} + m\mu_{\mathrm{W}}^{(\mathrm{ex})})$$

$$- \ln[K^{(0)}(C_{n,m})\rho_{\mathrm{W}}{}^m] + \beta\mu_{\mathrm{X}\cdot\mathrm{L}_n\mathrm{W}_m}^{(\mathrm{ex})} + \ln p_{\mathrm{X}}(C_{n,m})$$

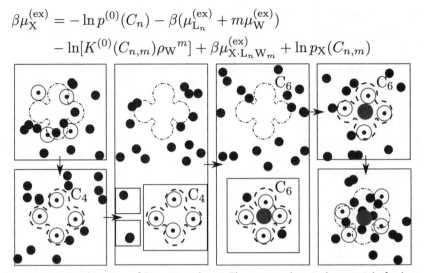

Figure 4.14 Combination of Figs 4.8 and 4.10. The excess chemical potential of solute X (red sphere) described by QCT as an ideal *solute binding* (third arrow) within a site formed from *m* solvent molecules, W (solid blue spheres), and *n* ligands, L (open spheres) uncoupled from the environment. C_n ensures that the ligands form the required coordination; $C_{n,m}$ constrains locations of the solvent and ligands forming the complex. As in Fig. 4.10, $\mu_{\mathrm{L}_n}^{(\mathrm{ex})}$ and $\mu_{\mathrm{W}_n}^{(\mathrm{ex})}$ represent the free energy for *removing*, and $\mu_{\mathrm{X}\cdot\mathrm{L}_n\mathrm{W}_m}^{(\mathrm{ex})}$ for *restoring*, interactions with the outer environment. The first and last terms represent ligand *binding site formation* without solute, and *constraint release*, as in Fig. 4.8. This follows steps A → B → C → D → E → F of cycle 2 in Ref. [113]. For interpretation of the references to color in this figure legend, the reader is referred to the online version of this book.

that the middle three components of Fig. 4.14 correspond to the *solute binding* component $(\ln\langle e^{-\beta\varepsilon}|C_n\rangle)$ of Fig. 4.8.

To illustrate application of Fig. 4.14, we return to the pK_{a} shift of the catalytic water at the Zn^{2+} binding site described in Chapter 4, Section 1.1. Mutations to the active site of carbonic anhydrase shift the acid dissociation constant of the catalytic water (see Fig. 4.1). Specifically, mutations of the zinc-coordinating histidines to aspartic acid residues raise the pK_{a}, making the catalytic water less acidic and thus hindering uptake of carbon dioxide by the enzyme. How do we understand these shifts in pK_{a} and especially the observation that the shifts vary depending on which of the coordinating histidines is mutated?

First, it must be understood that the pK_{a} is a local, structural property of a proton-binding site. Only then does it make sense to speak of the pK_{a} of a residue or a functional group of a molecule, excluding the possibility for the proton to associate with a different group. The pK_a comes from the reverse of the acid *association* reaction:

$$H^+ + A^- (C_n) \rightleftharpoons HA(C_n), \tag{4.10}$$

under a condition, C_n, on the structure of the conjugate base, A^-, relative to the fixed proton location. Note that the association direction is used here in order to show the direct parallel with Fig. 4.8 for solute $X = H^+$.

The free energy released when a proton is added to a point in solution can be decomposed using QCT (Fig. 4.8),

$$\mu_{H^+}^{(ex)} = k_B T \ln\langle e^{-\beta \varepsilon_{H^+}} | C_n \rangle + k_B T \ln \frac{p_{H^+}(C_n)}{p^{(0)}(C_n)}, \tag{4.11}$$

to show the trade-off between the likelihood for protonation of an arbitrary binding site $\dfrac{p_{H^+}(C_n)}{p^{(0)}(C_n)}$ and the site's absolute proton-binding affinity, $k_B T \ln\langle e^{-\beta \varepsilon_{H^+}} | C_n \rangle$.

Differences in the chemical equilibria of Eqn (4.10) between alternate binding sites give relative pK_a values,

$$k_B T \ln(10)\Delta pK_a = -k_B T \Delta\ln\langle e^{-\beta \varepsilon_{H^+}} | C_n \rangle. \tag{4.12}$$

These are independent of the choice of standard state.

To analyze shifts in relative pK_a with mutations to the catalytically active site of carbonic anhydrase, we modeled the active site as zinc ion and molecules representing its four (4) coordinating ligands: water plus imidazoles for the three (3) histidine residues (H119, H94, H96) or acetate for each histidine mutated to aspartic acid (H to D). Applying QCT (Fig. 4.14), we treated the inner-shell active site quantum mechanically because of the highly polar local environment around zinc. A dielectric continuum with $\varepsilon = 40$ described the outer-shell solvation environment. This effective dielectric constant represents the collective electrostatic contributions from both the surrounding protein and the solvent [145]. We assumed that the surrounding environment permits structural rearrangement of the active site architecture to low-energy configurations. The energy-minimized structure of the wild-type carbonic anhydrase supports this assumption as it deviates only modestly (RMSD=1.3 Å) from the crystal structure [146].

Table 4.1 shows components from the QCT free energy decomposition (Fig. 4.14). The chemical dissociation of H^+ from zinc-bound water in the

Table 4.1 Free energy components of pK_a calculations for wild type and mutant CA II and shifts in dipole moment of the catalytic water [146].

	$\ln[K^{(0)}(C_n)]^a$	$-\Delta\mu^{(ex)}_{H^+\cdot Zn^{2+}L_3OH^-}$ [b]	$-\Delta\ln\langle e^{-\beta\varepsilon_{H^+}}|C_n\rangle^c$	ΔpK_a^d	Δm^f
WT	181.93	103.48	—	—	—
H119D	262.94	25.26	2.79	$2.03(1.8)^e$	−0.03
H94D	261.39	27.84	3.82	$2.78(2.8)^e$	0.60
H96D	261.05	29.19	4.83	3.52	0.66

$X = H^+$; C_n is the active site with $n = 3$ coordinating ligands (L) and a water. Free energies are given in kcal/mol and dipole moment, m, in Debye:
[a] proton dissociation free energy (*chemical contribution* in vapor phase);
[b] *outer environment* contribution, $\Delta\mu^{(ex)}_{H^+\cdot Zn^{2+}L_3OH^-} = \mu^{(ex)}_{H^+\cdot Zn^{2+}L_3OH^-} - \mu^{(ex)}_{Zn^{2+}L_3OH^-}$;
[c] total deprotonation free energy difference between mutant and native enzymes;
[d] relative changes in pK_a of the active site with mutations;
[e] experimental data from Ref. [11] listed in parentheses;
[f] change in magnitude of the dipole moment from protonated to deprotonated water follows the trend in pK_a shifts with mutations.

active site uncoupled from the surrounding environment is less favorable in the negatively charged mutants compared to the wild type, while the solvation contribution from *outer-shell* interactions is more favorable.

The large and opposing changes of the chemical and outer-shell free energy components mostly cancel, producing small unfavorable shifts in acid dissociation constant to higher pK_a values for each mutant. Achieving the correct balancing of these opposing contributions is challenging. Future work will benefit from more careful analysis of the outer-shell contributions and more accurate estimation of contributions from changes in binding site conformation. Nevertheless, the calculations presented here compare favorably with experimental data available for two of the mutants. They indicate that single mutations of the active site histidines (H119, H94, H96) to aspartic acid will shift the zinc-bound water pK_a upward by 2.0, 2.8, and 3.5 pK_a units.

Analysis of the active site structure and electron distribution demonstrates that the pK_a changes correlate with the change in dipole moment of the zinc-bound water as it converts between protonated and deprotonated forms. This change in dipole moment varies with location of the mutation because of the natural asymmetry in the binding site combined with changes to the coordination structure around zinc from 4- to 5-coordinate, depending on location of the mutation.

Ion-selective chelators and macromolecules present computationally challenging targets for mechanistic studies, in general, due to the strong *chemical* interactions. We note here that a series of recent reviews has converged on the importance of local structure, in addition to any environmental bias of that local structure, in modeling and understanding this topic [147]. Structural decompositions of the free energy of binding-site formation [113] thus provide a central focal point for comparing minimalistic binding-site and large-scale, nonequilibrium [148] models of selectivity and transport.

4. STATISTICAL MECHANICAL MODELING OF IONIC CORRELATIONS

The role of statistical mechanical theory in understanding electrolyte solutions at a molecular level has shifted substantially with the availability of molecular simulation tools. Simulation calculations often bypass dominating obstacles of nonsimulation statistical mechanical theory without comment or resolution. This maneuver sometimes identifies obstacles that can be overcome by brute force without addressing simpler molecular mechanisms. Nevertheless, some of those theoretical obstacles do reflect our conceptual understanding. Those concepts deserve eventual resolution even if alternative brute force computations are widely available. Principal questions involve the effects of inter-ionic correlations.

We take two such conceptual points to underpin the discussion that follows. The first of these is the coarse-graining that eliminates the solvent molecules from direct consideration when treating electrolyte solutions. The second conceptual point is a reconsideration of the PB equation as a basis of the theory of electrolyte solutions. Both of these concepts dominate contemporary thinking about the physical chemistry of aqueous solutions in the context of biomolecular science.

4.1. Primitive, or McMillan–Mayer, Models of Electrolyte Solutions

The issue for the first of these concepts is that it is common to begin by considering ions in solutions with the solvent replaced by a uniform dielectric medium. In a primitive case, for example, hard-spherical ions are considered, the solvent is not actively present, and when the ion hard spheres do not overlap, the interactions between ions is assumed to be $q_i q_j / 4\pi\varepsilon r$, where q_i is the formal charge on the ith ion, $\varepsilon/\varepsilon_0$ is the dielectric constant of the solvent, and r is the separation of the charges of the ions. Such models are typically called "primitive models," literally. Such models are not justified by direct molecular-scale observation of the solvent, i.e. the solvent is not actually a dielectric continuum. The justification must be more subtle and that is the target of statistical mechanical theory.

The justifications sought are often intuitive [149], and sometimes directly empirical as in fitting modeled thermodynamic properties to data [150]. But the theory for elimination of the solvent in this way has been conclusively considered: it is the McMillan-Mayer (MM) theory [151,152,153], and it is the pinnacle of the theory of coarse-graining for the statistical mechanics of solutions. "Primitive model" is then synonymous [153,154] with "McMillan-Mayer model," and we can use those names interchangeably.

The foremost feature of MM theory is that the solvent coordinates are fully *integrated-out*. The statistical mechanical problem that results from MM analysis treats the solute molecules (ions) only, but with effective interactions that are formally fully specified. Those effective interactions are complicated [155,156]. It can be argued that no sacrifice of molecular realism is implied by MM theory. But cataloging of the multi-body potentials implied by a literal MM approach is prohibitively difficult [157]. Use of MM theory to construct specifically a primitive model for a system of specific experimental interest has been limited [155,156].

The integrating-out does accomplish a coarse-graining, and from a thermodynamic point of view, the coordinates eliminated are the solvent or solvent moles or numbers (n_S) of molecules. Those coordinates are obvious from the thermodynamic specification of the problem. This is a favored case for theoretical considerations because identification of degrees of freedom to eliminate will never be less arbitrary than that. The effective interactions that result depend on the thermodynamic state of the solvent, of course, and specifically on the chemical potential of the solvent, μ_S. This is consistent with the physical picture of MM analysis that the system under study can be viewed as in osmotic equilibrium with pure solvent at a specific chemical potential.

With the statistical mechanical community focusing on primitive models, an enormous literature (illustrative examples: [107,153,158–160]) of sophisticated theory became available to treat inter-ion correlations. Historic PB equation concepts faded, and most of the intervening statistical mechanical literature does not even reference "PB." Despite that statistical mechanical progress, biophysical modeling that treats electrolyte solutions has been overwhelmingly dominated by applications of PB equation approaches. Essentially, all biomolecular simulation packages anticipate application of PB calculations to the biomolecular structures considered or provide those tools explicitly.

We conclude this subsection with the view that when traditional Debye–Hückel or PB treatments of electrolyte solutions fail to agree accurately with experiment, the first concern is the accuracy of the primitive (or MM) model. Contrariwise, when the traditional treatments do accurately agree with experiment, the first challenge for molecular understanding is to explain why the assumed primitive model is indeed satisfactory.

4.2. PB Approaches

On the one hand, the PB theory was recognized in the 1930s to be thermodynamically inconsistent beyond the Debye–Hückel (DH) limit [161,162]. That is the reason that presently a broad statistical mechanical community does not consider it further. On the other hand, the biomolecular modeling researchers focus almost exclusively on PB, and it is important to recognize why. The first reason is that the shapes of the biomolecules are essential to those problems and the shapes are complicated. Application of PB theory to complicated shaped molecules is conceptually natural. The second reason is that PB theory produces electrostatic potentials and electric field lines that facilitate graphic communication of the results for complicated applications. Thus, PB theories are built upon basic concepts that support physically appealing molecular intuition.

It may be the last of these points—access to physical intuition—that is the most important. It is thus appropriate that the statistical mechanics community return to PB concepts with the maturation of the statistical mechanical theory of these problems. Recent developments of the local molecular field (LMF) theory [163–165], and the observations below, address this point. Therefore, it is helpful here to give a high-level description of the historic PB approach. We will pursue that description by considering inter-ionic potentials of the average forces (or "potentials of mean force" or "pmfs") between an ion-pair $\alpha\gamma$:

$$- \beta w_{\alpha\gamma}(r) = \ln \, g_{\alpha\gamma}(r) \qquad (4.13)$$

with $g_{\alpha\gamma}(r)$ the usual radial distribution function for $\alpha\gamma$ ion pair. Direct modeling of $w_{\alpha\gamma}(r)$ can be taken to be defining a characteristic of PB approaches. In the historic case, the inter-ionic interactions beyond necessary excluded volume interactions are solely Coulomb interactions. Then, it is assumed that $w_{\alpha\gamma}(r) = q_\alpha \psi_\gamma(r)$, where the excluded volume interactions vanish, and the basic approximation is [161,162]

$$\varepsilon \nabla^2 \psi_\gamma (r) \approx -q_\gamma \delta(\vec{r}) - \sum_\eta \rho_\eta q_\eta e^{-\beta q_\eta \psi_\gamma(r)} \qquad (4.14)$$

with appropriate boundary conditions. This is the PB equation. Here, $\psi_\gamma(r)$ is the electrostatic potential induced by a charge q_γ at the origin, $\vec{r} = \vec{0}$. For ion concentrations $\rho_\eta \sim 0$, linearization of the right side of Eqn (4.14) and acknowledgment that $\sum_\eta \rho_\eta q_\eta = 0$ then identifies the combination of parameters $\kappa^2 = \beta \sum_\eta \rho_\eta q_\eta / \varepsilon$. Dimensionally, κ^{-1} is a length and is referred to as the "Debye screening length." Important for the discussion that follows is that κ is integral to the DH contribution to the excess chemical potential due to inter-ionic correlations; for a 1:1 electrolyte in the limiting DH ($\rho_\alpha \sim 0$) circumstance is

$$\left(\frac{1}{2} \right) \left(\mu_+^{(ex)} + \mu_-^{(ex)} \right) \sim -\left(\frac{1}{2} \right) \frac{e^2 \kappa}{4\pi\varepsilon}, \qquad (4.15)$$

a particularly simple combination of the parameters for the present problem.

With this background, we can highlight recent statistical mechanical contributions advancing PB approaches. We start with the obvious point that treatment of actual systems, in contrast to academic models, must consider nonionic interactions, e.g. van der Waals interactions with excluded volume and dispersion contributions.

The LMF theory [163–165] does expand the scope of PB to include interactions other than obvious ionic interactions. But the LMF theory sheds light on the historic PB theory initially when applied to the same ionic interaction models as the historic theory. The first step on the LMF conceptual path is then to separate the ionic interactions into a long-ranged part (perturbation interactions), and a remainder (reference interactions). Long-ranged, ionic interactions are the first difficulty of ionic solutions, and the perturbation interactions must match those ionic interactions at longest range. The possibility that has been studied so far can be defined as the electrostatic potential of a diffuse, but normalized charge distribution such as $\phi(r; \sigma)$:

$$\varepsilon \nabla^2 \psi(r; \sigma) = -q\phi(r; \sigma). \qquad (4.16)$$

σ is a length parameter that characterizes the width of the distribution and $\phi(r; \sigma) = e^{-r^2/2\sigma^2}/\sqrt[3]{2\pi\sigma^2}$ is the case studied in detail so far. The model Eqn (4.16) behaves as a simple Coulomb field for $r \gg \sigma$ but is smooth with limited variation at small distances. The reference interactions must complement these perturbation interactions.

The reference interactions, being short ranged, include whatever excluded volume interactions are naturally present with MM models. It is assumed that the reference system problem will be solved by brute force, i.e. by molecular simulation. Then as a second conceptual step, the LMF theory focuses on treatment of the long-ranged perturbation.

From this perspective, *fixing* the PB theory means solving in turn *two* problems, firstly the reference system problem by brute force. Let us call that result $w_{\alpha\gamma}^{(0)}(r; \sigma)$. Then, secondly, a PB problem for inclusion of the effects of the perturbation is required. That second PB step evaluates the effect, $\overline{\psi}_\gamma(r; \sigma)$ of the perturbative interactions according to

$$\varepsilon \nabla^2 \overline{\psi}_\gamma(r; \sigma) = -q_\gamma \phi(r; \sigma) - \int \phi\left(|\vec{r} - \vec{r}\,'|; \sigma\right)$$

$$\times \sum_\eta \rho_\eta q_\eta e^{-\beta\left(w_{\eta\gamma}^{(0)}(\vec{r}\,';\sigma) + q_\eta \overline{\psi}_\gamma(\vec{r}\,';\sigma)\right)} d\vec{r}\,'. \qquad (4.17)$$

The smearing function $\phi(r; \sigma)$ enforces the smoothness of the effects of the perturbation, $\overline{\psi}_\gamma(r; \sigma)$. Equation (4.17) is expected to be an accurate representation of the effects of the perturbation provided that σ is sufficiently large that the perturbation is slowly varying. Specific applications must establish what are suitable values of σ that preserve accuracy of the theory while achieving reasonable computational efficiency in treating the reference system.

An interesting consequence of the LMF development is that, with this organization of the interactions and sufficiently large σ, historic theories such as the random phase approximation (RPA) are rescued in that they become helpfully accurate for treatment of effects of perturbation [165]. The physical conclusion is evidently that electrostatic interactions at *short range* $(r \leq \sigma)$ are dominating contributions for the typical solution models. Including those short-ranged electrostatic contributions in the reference interactions to be treated by brute force is a crucial step toward making the effects of the perturbations simple. In the context of QCT, this is particularly interesting because RPA treatments imply a Gaussian (normal) distribution of perturbation contributions to binding energies [166]; this issue will be discussed further below.

4.3. Binding Energy Distributions

Direct modeling of the pmfs $w_{\alpha\gamma}(r)$ can be based on the PDT also. There is no formal difficulty in considering not-necessarily-spherical molecular ions, so we will initially denote the general configuration of an $\alpha\gamma$ ion pair by (1,2) and the desired pmf by $w_{\alpha\gamma}$ (1,2). A basic formula is [3,167,168]

$$e^{-\beta w_{\eta\nu}(1,2)} = e^{-\beta u_{\eta\nu}(1,2)} \left[\frac{\left\langle\left\langle e^{-\beta \Delta U_{\eta\nu}^{(2)}} \Big| 1,2 \right\rangle\right\rangle_0}{\left\langle\left\langle e^{-\beta \Delta U_{\eta}^{(1)}} \right\rangle\right\rangle_0 \left\langle\left\langle e^{-\beta \Delta U_{\nu}^{(1)}} \right\rangle\right\rangle_0} \right]. \qquad (4.18)$$

The interaction potential energy for the isolated ion pair is $u_{\eta\nu}$ (1, 2). The denominator factors are the conventional potential distribution expressions of the excess chemical potentials [Eqn (4.2)]. Though it is not generally necessary, these formulae assume that a uniform-fluid system is considered. The binding energy of a test ion is denoted by $\Delta U_{\nu}^{(1)} = U(N+1) - U(N) - U_{\nu}(1)$. The notation $\left\langle\left\langle e^{-\beta \Delta U_{\eta\nu}^{(2)}} \big| 1,2 \right\rangle\right\rangle_0$ is a natural generalization of the PDT to consider a test pair of molecules (1,2), and the brackets $\left\langle\left\langle \ldots |1,2 \right\rangle\right\rangle_0$ signify a conditional expectation, conditional on location of the ion pair. This uses $\Delta U_{\eta\nu}^{(2)} = U(N+2) - U(N) - u_{\eta\nu}$ (1, 2) and does not assume pair-decomposable interactions. The subscript 0 indicates that this is the *forward* use of the PDT; the averaging takes no account of the interactions between medium and the test molecule pair, though $\Delta U_{\eta\nu}^{(2)}$ of course does express those interactions.

If a calculation will be carried out with the ions actually present, exerting interactions that influence the surrounding medium, then the appropriate formula adopts the inverse form

$$e^{-\beta w_{\eta\nu}(1,2)} = e^{-\beta u_{\eta\nu}(1,2)} \left[\frac{\left\langle e^{\beta \Delta U_{\eta}^{(1)}} \right\rangle \left\langle e^{\beta \Delta U_{\nu}^{(1)}} \right\rangle}{\left\langle e^{\beta \Delta U_{\eta\nu}^{(2)}} \big| 1,2 \right\rangle} \right]. \qquad (4.19)$$

In this case, the brackets indicate the thermal average with interactions between medium and test pair fully operational.

The idea of analyzing pair interactions on the basis of solvation free energies of specified pairs is widely appreciated on a specific but informal basis. Refs. [169,170] give examples. Here, we consider these ideas more formally in order to lay a basis of consideration of ionic correlations.

The point of interest for the present discussion is that the averages indicated in these equations can be simple when long-ranged interactions are considered, as in Section 4.1. To exemplify this point, let us consider the DH theory resulting in Eqn (4.15). We write

$$e^{-\beta\mu_\eta^{(ex)}} = \int e^{-\beta\varepsilon} P_\eta^{(0)}(\varepsilon) d\varepsilon, \tag{4.20}$$

and investigate the distribution $P_\eta^{(0)}(\varepsilon) = \langle\langle\delta(\varepsilon - \Delta U_\eta^{(1)})\rangle\rangle_0$ for the DH theory of a restricted primitive model. We adopt the notation that $\varepsilon = q_\eta\psi$ and write [171]

$$e^{-\beta q_\eta^2\kappa/8\pi\varepsilon} = \int e^{-\beta q_\eta\psi} P_\eta^{(0)}(\psi) d\psi \tag{4.21}$$

with $\int P_\eta^{(0)}(\psi)d\psi = 1$. Because of the quadratic q_η dependence on the left side, it is clear that

$$P_\eta^{(0)}(\psi) = \frac{\exp\left\{-\psi^2/2\langle\psi^2\rangle_0\right\}}{\sqrt{2\pi\langle\psi^2\rangle_0}}, \tag{4.22}$$

a zero-mean Gaussian (normal) distribution with variance $\langle\psi^2\rangle_0 = \kappa/4\pi\varepsilon\beta$ [171]. Substitution of this result into Eqn (4.21) specifically confirms that this is correct.

When ion-pair pmfs are known in detail, we can use Eqn (4.18) to examine the joint distribution of the electric potentials at two neighboring points separated by the distance r. The Debye–Hückel result $w_{\eta\nu}(r) = q_\eta q_\nu e^{-\kappa r}/4\pi\varepsilon r$ provides a specific example [168,172,173]. To proceed systematically, we note that the substitution $\beta q_\eta = ik$ casts Eqn (4.21) as a Fourier integral transform. The corresponding formulation for the joint two-point distribution is

$$\int e^{-ik_\alpha\psi_\alpha}e^{-ik_\eta\psi_\eta} P^{(0)}(\psi_\alpha, \psi_\gamma) d\psi_\alpha d\psi_\gamma$$

$$= \exp\left\{-\left(k_\alpha^2 - 2\phi(\kappa r)k_\alpha k_\gamma + k_\gamma^2\right)\langle\psi^2\rangle_0/2\right\}, \tag{4.23}$$

with $\phi(\kappa r) = (1 - e^{-\kappa r})/4\pi\varepsilon\kappa r$, using Eqn (4.18). Methodically carrying out the inverse Fourier transform produces

$$P^{(0)}\left(\psi_\alpha, \psi_\gamma\right) = \frac{\exp\left\{-\dfrac{\psi_\alpha^2 - 2\psi_\alpha\psi_\gamma\phi\left(\kappa r\right) + \psi_\gamma^2}{2\langle\psi^2\rangle_0\left(1 - \phi(\kappa r)^2\right)}\right\}}{2\pi\langle\psi^2\rangle_0\sqrt{1 - \phi(\kappa r)^2}}. \qquad (4.24)$$

This result was known previously [168,172,173] and reinforces Eqn (4.18). The interesting observation [172] is that $\phi(r) \sim 1/r$, so that electric potentials are positively correlated at large distances. This can be appreciated physically by the following argument. For actual ions, a counter-ion cloud shields the potential so that the net effect is short ranged. But the potential due solely to the distinguished ion is long ranged, so in compensation, the potential due to the shielding counter-ion cloud is long ranged also. Together, the combined effect is short ranged.

This detailed calculation here depends on the displayed q-dependence of the results for test charges and pairs. Our calculation also reinforces the point that more sophisticated theories built around diagrammatic chain, and ring summations [107,166] will result in Gaussian distributions also with substitution of the appropriate different results for the moments required.

We include these general results in the present review because the preceding discussions have emphasized that the QCT development (Fig. 4.2) has shed new light on the statistical thermodynamics of realistically modeled solutions. We anticipate that similar analyses of the formulae such as Eqn (4.19) will be similarly helpful, though clearly a serious challenge.

5. CONCLUSIONS AND OUTLOOK

The QCT division of the absolute free energy of solvation builds the foundation for molecular theories incorporating local and long-ranged molecular structure. Its successes include accurate statistical mechanical models of hydrophobic hydration, connection to the local solvent structure in ion and polar molecule solvation, and a statistical mechanical proposal for implicit/explicit solvent models.

Applied to realistic cases where van der Waals interactions dominate the general mixture of solute–solvent interactions, QCT does not require an *a priori* assessment of solute cavity size. This demonstrates that the present QCT represents a significant generalization of van der Waals approaches.

Applications of QCT have given molecular-scale information on the solvation process. Metal cation solvation structures in water can be well described by a few strongly bound waters, while longer-ranged interactions follow a simple dielectric model. Halide anions have less well-defined structures, which show some degree of anisotropy. This anisotropy requires the use of more careful modeling of the local solvent structure around the solute. Anisotropic distributions of solvent around ion pairs [174,175] can have appreciable influences on solvation free energies and potentials of mean force.

For such hydrophilic molecules, Gaussian models for the solute–solvent interaction energy distribution have been found to work well [86,102]. These have a direct connection to linear response theory for solvent polarization. Accurate statistical mechanical models of solvent packing plus a correction for dispersion interactions leads to quantitative models for solvent structure and solubility of hydrophobic solutes such as methane and CF_4. The packing plus long-ranged electrostatics and dispersion model applies well to intermediate, nonpolar solutes such as CO_2.

This general view of solvation develops from considering individual structural states. QCT theory does this by balancing contributions from ligand chemistry, coordination number, cavity shape, flexibility, and characteristics of the external environment. Each of these influences ion selectivity, whether in liquids or macromolecular binding sites. The contributions can be computed by classifying structures sampled during thermodynamic perturbation calculations. Applications to proteins highlighted an *architectural control mechanism* of selective ion binding, where interactions with the surrounding environment constrained ligand composition and configuration. Here, ion selectivity may be tuned by changing the surroundings, rather than the binding site chemical motif. Application to carbonic anhdyrase facilitated analysis of the structural and chemical basis of pK_a shifts that favor CO_2 release over uptake in mutated active sites.

Because of its statistical mechanical starting point, the QCT decomposition provides a supporting basis for other methods for local structural simulation. It can provide a foundation for analysis of the solution thermodynamics of energy decomposition methods such as self-consistent reaction field theory [176,177] or mixed QM/MM Hamiltonian models. Reviews cover such models in depth [178]. The simple idea of modeling the stochastic solvent distribution near a surface in the framework of QCT [1] has also been developed and found to describe deviations in the osmotic virial coefficient from colloidal models of protein–protein interactions [179].

Developments on aqueous solution interfaces and specific ion effects provide another glimpse of what the future may hold for these methods. In Ref. [180], a mean-field model treats long-ranged dispersion and electrostatic forces between atoms. Molecular-scale solvent structure is tied to short-ranged interactions. Thus, solvent structure may be computed efficiently using algorithms that scale linearly with the number of available processors. The long-ranged field can be estimated using lower-resolution theories, as well as iteratively refined.

The error function decomposition of electrostatic forces as used in Ref. [180] has also given structural insight into the origins of the Hofmeister series [181]. The entropy of ionic solvation was split into local plus long-ranged parts using an electrostatic cutoff at the first minimum of the ion-oxygen radial distribution function. The long-ranged solvation entropy was negative for all ions studied, in line with the Born model, but the local solvation entropy increased from kosmotropic (structure-making) through chaotropic (structure-breaking) ions in a Hofmeister series. This is the first simulation evidence for the conjecture that the disorder in the ion's local solvation shell contributes quantitatively to Hofmeister effects [182].

As a statistical mechanical theory of solvation, QCT sits at the interface between the fields of physical chemistry, biophysics, colloid science, and simulations of organic and inorganic small-molecule solution reactivity. For physical chemistry of solutions, extensions of classical integral theories to deal with molecular shape and nuances of solvation structure are required. For biophysics, methods for estimating solvation contributions to conformational free energies and potentials of mean force between proteins, membranes, and highly charged entities such as heparin and DNA are desperately needed. Colloid science is presently searching for new methods to understand deviations of surface forces from DLVO (Derjaguin, Landau, Verwey, Overbeek) theory due to local solvent structure and dispersion interactions [18]. Simulation and analysis capabilities for molecular reactions and thermochemistry in solution are currently generating raw data from quantum and molecular mechanical models that could be usefully employed in statistical mechanical theories. Force field parameterization provides a prime example of where high-level theories can aid in identifying features necessary for constructing physical models. What is required is more collaboration between these areas to make solid progress on transferrable models of solvation.

ACKNOWLEDGMENTS

Part of this material is based upon work supported by Sandia's Laboratory Directed Research and Development and part is based upon work supported by the National Science

Foundation under the NSF EPSCoR Cooperative Agreement No. EPS-1003897, with additional support from the Louisiana Board of Regents. Sandia National Laboratories is a multiprogram laboratory managed and operated by Sandia Corporation, a wholly owned subsidiary of Lockheed Martin Corporation, for the U.S. Department of Energy's National Nuclear Security Administration under contract DE-AC04-94AL85000. LRP thanks Dilip Asthagiri for helpful discussions and collaborations in development of the molecular quasi-chemical approach. SBR thanks Sameer Varma for collaborations in development and applications of QCT to protein-binding sites.

REFERENCES

[1] Pratt, L. R.; Rempe, S. B. Quasi-Chemical Theory and Implicit Solvent Models for Simulations In *Simulation and Theory of Electrostatic Interactions in Solution,* Vol. 492 of AIP Conference Proceedings; Hummer, G., Pratt, L. R., Eds.; AIP Press: New York, 1999; pp 172–201.

[2] Paulaitis, M. E.; Pratt, L. R. Hydration Theory for Molecular Biophysics; Adv. Prot. Chem. **2002**, *62*, 283–310.

[3] Beck, T. L.; Paulaitis, M. E.; Pratt, L. R. *The Potential Distribution Theorem and Models of Molecular Solutions*; Cambridge: New York, 2006.

[4] Pratt, L. R.; Asthagiri, D. Potential Distribution Methods and Free Energy Models of Molecular Solutions In *Free Energy Calculations. Theory and Applications in Chemistry and Biology,* Vol. 86 of Springer Series in Chemical Physics; New York, 2007.

[5] Asthagiri, D.; Dixit, P. D.; Merchant, S.; Paulaitis, M. E.; Pratt, L. R.; Rempe, S. B.; Varma, S. Ion Selectivity from Local Configurations of Ligands in Solutions and Ion Channels; Chem. Phys. Lett. **2010**, *485* (1–3), 1–7.

[6] Chempath, S.; Pratt, L. Distribution of Binding Energies of a Water Molecule in the Water Liquid–Vapor Interface; J. Phys. Chem. B **2009**, *113*, 4147–4151.

[7] Lindskog, S. The Catalytic Mechanism of Carbonic Anhydrase; Proc. Natl. Acad. Sci. USA **1973**, *70*, 4.

[8] Lindskog, S. Structure and Mechanism of Carbonic Anhdyrase; Pharm. Therapeut. **1997**, *74*, 1–20.

[9] Mohammed, O. F.; Pines, D.; Dreyer, J.; Pines, E.; Nibbering, E. T. J. Sequential Proton Transfer Through Water Bridges in Acid–Base Reactions; Science **2005**, *310* (5745), 83–86, doi:10.1126/science.1117756

[10] Riccardi, D.; Konig, P.; Prat-Resina, X.; Yu, H.; Elstner, M.; Frauenheim, T.; Cui, Q. "Proton holes" in Long-Range Proton Transfer Reactions in Solution and Enzymes: A Theoretical Analysis; J. Am. Chem. Soc. **2006**, *128* (50), 16302–16311.

[11] Kiefer, L. L.; Fierke, C. A. Functional Characterization of Human Carbonic Anhydrase II: Variants with Altered Zinc Binding Sites; Biochemistry **1994**, *33* (51), 15233–15240.

[12] March, J. *Advanced Organic Chemistry*, 3rd ed.; John Wiley and Sons Inc: Hoboken, NJ, 1985.

[13] Obst, S.; Bradaczek, H. Molecular Dynamics Simulations of Zinc Ions in Water Using CHARMM; J. Mol. Model **1997**, *3*, 224–232.

[14] Kuzmin, A.; Obst, S.; Purans, J. X-ray Absorption Spectroscopy and Molecular Dynamics Studies of Zn^{2+} Hydration in Aqueous Solutions; J. Phys. Condens. Matter **1999**, *9* (46), 10065–10078.

[15] Asthagiri, D.; Pratt, L. R.; Paulaitis, M. E.; Rempe, S. B. Hydration Structure and Free Energy of Biomolecularly Specific Aqueous Dications, Including Zn^{2+} and First Transition Row Metals; J. Am. Chem. Soc. **2004**, *126* (4), 1285–1289.

[16] Ashbaugh, H. S. Convergence of Molecular and Macroscopic Continuum Descriptions of Ion Hydration; J. Phys. Chem. B **2000**, *104* (31), 7235–7238.

[17] Friedman, H. L.; Krishnan, C. V. Thermodynamics of Ion Hydration In. *Water: A Comprehensive Treatise*; Franks, F., Ed.; Plenum Press: New York, 1973; Vol. 3, pp 1–118.

[18] Parsons, D. F.; Bostrom, M.; Nostro, P. L.; Ninham, B. W. Hofmeister Effects: Interplay of Hydration, Nonelectrostatic Potentials, and Ion Size, Phys. Chem. Chem. Phys. **2011**, *13* (27), 12352–12367.

[19] Zhu, P.; You, X.; Pratt, L. R.; Papadopoulos, K. Generalizations of the Fuoss Approximation for Ion Pairing; J. Chem. Phys. **2011**, *134*, 054502.

[20] Manning, G. S.; Ray, J. Counterion Condensation Revisited; J. Biomol. Struct. Dyn. **1998**, *16* (2), 461–476.

[21] Barbosa, M. C.; Deserno, M.; Holm, C. A Stable Local Density Functional Approach to Ion–Ion Correlations; Europhys. Lett. (EPL) **2000**, *52* (1), 80–86.

[22] Lodi, L.; Tennyson, J. Theoretical Methods for Small-Molecule Ro-Vibrational Spectroscopy; J. Phys. B At. Mol. Opt. Phys. **2010**, *43* (13), 133001.

[23] Rempe, S. B.; Watts, R.O. The Convergence Properties of Hindered Rotor Energy Levels; Chem. Phys. Lett. **1997**, *269*, 455–463.

[24] Rempe, S. B.; Watts, R. O. The Exact Quantum Mechanical Kinetic Energy Operator in Internal Coordinates for Vibration of a Hexatomic Molecule; J. Chem. Phys. **1998**, *108*, 10084.

[25] Varma, S.; Rempe, S. B. Importance of Multi-Body Effects in Ion Binding and Selectivity; Biophys. J. **2010**, *99*, 3394–3401.

[26] Zhao, Z.; Rogers, D. M.; Beck, T. L. Polarization and Charge Transfer in the Hydration of Chloride Ions; J. Chem. Phys. **2010**, *132* (1), 014502.

[27] Bauer, B. A.; Lucas, T. R.; Krishtal, A.; Alsenoy, C. V.; Patel, S. Variation of Ion Polarizability from Vacuum to Hydration: Insights from Hirshfeld Partitioning; J. Phys. Chem. A **2010**, *114* (34), 8984–8992.

[28] McGrath, M. J.; Siepmann, J. I.; Kuo, I. W.; Mundy, C. J.; VandeVondele, J.; Hutter, J.; Mohamed, F.; Krack, M. *J. Phys. Chem. A* **2006**, *110*, 640.

[29] Beck, T. Quantum Contributions to free Energy Changes in Fluids In *Free Energy Calculations*, Vol. 86 of Springer Series in Chemical Physics; Springer: New York, 2007; pp 389–422.

[30] Schwegler, E.; Grossman, J. C.; Gygi, F.; Galli, G. Towards an Assessment of the Accuracy of Density Functional Theory for First Principles Simulations of Water II; J. Chem. Phys. **2004**, *121*, 5400–5409.

[31] Israelachvili, J. N. *Intermolecular and Surface Forces*, 3rd ed.; Academic Press: Waltham, MA, 2011.

[32] *Casimir Physics*. In Lecture Notes in Physics 834; Dalvit, D., Milonni, P., Roberts, D., da Rosa, F., Eds.; Springer: New York, 2011.

[33] Berne, B. J.; Thirumalai, D. On the Simulation of Quantum Systems: Path Integral Methods; Ann. Rev. Phys. Chem. **1986**, *37*, 401–424.

[34] Liu, A.; Beck, T. L. Determination of Excess Gibbs Free Energy of Quantum Mixtures by MC Path Integral Simulations; Mol. Phys. **1995**, *86*, 225–233.

[35] Fanourgakis, G. S.; Xantheas, S. S. Development of Transferable Interaction Potentials for Water. V. Extension of the Flexible, Polarizable, Thole-type Model Potential (TTM3-F, v. 3.0) to Describe the Vibrational Spectra of Water Clusters and Liquid Water; J. Chem. Phys. **2008**, *128* (7), 074506.

[36] Auffinger, P.; Cheatham, T. E.; Vaiana, A. C. Spontaneous Formation of KCl Aggregates in Biomolecular Simulations: A Force Field Issue? J. Chem. Theory Comput. **2007**, *3* (5), 1851–1859.

[37] Guárdia, E.; Skarmoutsos, I.; Masia, M. On Ion and Molecular Polarization of Halides in Water; J. Chem. Theory Comput. **2009**, *5* (6), 1449–1453.

[38] Vreven, T.; Morokuma, K. Hybrid Methods: Oniom (QM: MM) and QM/MM In. *Ann. Rep. Comp. Chem.*; Spellmeyer, D., Ed.; Elsevier: 2006; Vol. 2, pp 35–51.

[39] Jensen, K. P.; Jorgensen, W. L. Halide, Ammonium, and Alkali Metal Ion Parameters for Modeling Aqueous Solutions; J. Chem. Theory Comput. **2006**, *2* (6), 1499–1509.

[40] Joung, I. S.; Cheatham, T. E., III. Determination of Alkali and Halide Monovalent Ion Parameters for Use in Explicitly Solvated Biomolecular Simulations; J. Phys. Chem. B **2008**, *112* (30), 9020–9041.

[41] Horinek, D.; Mamatkulov, S. I.; Netz, R. R. Rational Design of Ion Force Fields based on Thermodynamic Solvation Properties; J. Chem. Phys. **2009**, *130* (12), 124507.

[42] Jensen, K. P. Improved Interaction Potentials for Charged Residues in Proteins; J. Phys. Chem. B **2008**, *112* (6), 1820–1827.

[43] Ren, P.; Ponder, J. W. Polarizable Atomic Multipole Water Model for Molecular Mechanics Simulation; J. Phys. Chem. B **2003**, *107* (24), 5933–5947.

[44] Jorgensen, W. L.; Maxwell, D. S.; Tirado-Rives, J. Development and Testing of the OPLS All-Atom Force Field on Conformational Energetics and Properties of Organic Liquids; J. Am. Chem. Soc. **1996**, *118* (45), 11225–11236.

[45] Beglov, D.; Roux, B. Finite Representation of an Infinite Bulk System: Solvent Boundary Potential for Computer Simulations; J. Chem. Phys. **1994**, *100* (12), 9050–9063.

[46] Noskov, S. Y.; Roux, B. Control of Ion Selectivity in LeuT: Two Na^+ Binding Sites with Two Different Mechanisms; J. Mol. Biol. **2008**, *377* (3), 804–818.

[47] Bernèche, S.; Roux, B. Energetics of Ion Conduction Through the K^+ Channel; Nature **2001**, *414*, 73–77.

[48] Noskov, S. Y.; Bernèche, S.; Roux, B. Control of Ion Selectivity in Potassium Channels by Electrostatic and Dynamic Properties of Carbonyl Ligands; Nature **2004**, *431*, 830–834.

[49] Noskov, S. Y.; Roux, B. Importance of Hydration and Dynamics on the Selectivity of the KcsA and NaK Channels; J. Gen. Phys. **2007**, *129* (2), 135–143.

[50] Yu, H.; Mazzanti, C. L.; Whitfield, T. W.; Koeppe, R. E.; Andersen, O. S.; Roux, B. A Combined Experimental and Theoretical Study of Ion Solvation in Liquid N-Methylacetamide; J. Am. Chem. Soc. **2010**, *132* (31), 10847–10856.

[51] Varma, S.; Rogers, D. M.; Pratt, L. R.; Rempe, S. B. Design Principles for K^+ Selectivity in Membrane Transport; J. Gen. Physiol. **2011**, *137* (6), 479–488.

[52] Chen, A. A.; Pappu, R. V. Parameters of Monovalent ions in the AMBER-99 Forcefield: Assessment of Inaccuracies and Proposed Improvements; J. Phys. Chem. B **2007**, *111* (41), 11884–11887.

[53] Halgren, T. A. The Representation of van der Waals (vdW) Interactions in Molecular Mechanics Force Fields: Potential Form, Combination Rules, and vdW Parameters; J. Am. Chem. Soc. **1992**, *114* (20), 7827–7843.

[54] Weerasinghe, S.; Smith, P. E. A Kirkwood–Buff Derived Force Field for Sodium Chloride in Water; J. Chem. Phys. **2003**, *119* (21), 11342–11349.

[55] Luo, Y.; Roux, B. Simulation of Osmotic Pressure in Concentrated Aqueous Salt Solutions; J. Phys. Chem. Lett. **2009**, *1* (1), 183–189.

[56] Yoo, J.; Aksimentiev, A. Improved Parametrization of Li^+, Na^+, K^+, and Mg^{2+} Ions for All-Atom Molecular Dynamics Simulations of Nucleic Acid Systems; J. Phys. Chem. Lett. **2012**, *3* (1), 45–50.

[57] Hurst, GJ. B.; Fowler, P. W.; Stone, A. J.; Buckingham, A. D. Intermolecular Forces in van der Waals Dimers; Int. J. Quantum Chem. **1986**, *29* (5), 1223–1239.

[58] Jones, J. E. On the Determination of Molecular Fields. II. From the Equation of State of a Gas; Proc. Royal Soc. London A **1924**, *106* (738), 463–477.

[59] Wallqvist, A.; Ahlstroem, P.; Karlstroem, G. New Intermolecular Energy Calculation Scheme: Applications to Potential Surface and Liquid Properties of Water; J. Phys. Chem. **1990**, *94* (4), 1649–1656.

[60] Ichikawa, K.; Kameda, Y.; Yamaguchi, T.; Wakita, H.; Misawa, M. Neutron-Diffraction Investigation of the Intramolecular Structure of a Water Molecule in the Liquid Phase at High Temperatures; Mol. Phys. **1991**, *73* (1), 79–86.

[61] Krishtal, A.; Senet, P.; Yang, M.; Alsenoy, C. V. A Hirshfeld Partitioning of Polarizabilities of Water Clusters; J. Chem. Phys. **2006**, *125* (3), 034312.

[62] Jorgensen, W. L.; Tirado-Rives, J. Potential Energy Functions for Atomic-level Simulations of Water and Organic and Biomolecular Systems; Proc. Nat. Acad. Sci. USA **2005**, *102* (19), 6665–6670.

[63] Borgis, D.; Staib, A. A Semiempirical Quantum Polarization Model for Water; Chem. Phys. Lett. **1995**, *238* (13), 187–192.

[64] Lefohn, A. E.; Ovchinnikov, M.; Voth, G. A. A Multistate Empirical Valence Bond Approach to a Polarizable and Flexible Water Model; J. Phys. Chem. B **2001**, *105* (28), 6628–6637.

[65] Dang, L. X.; Chang, T. Molecular Dynamics Study of Water Clusters, Liquid, and Liquid–Vapor Interface of Water with Many-Body Potentials; J. Chem. Phys. **1997**, *106* (19), 8149–8159.

[66] Lamoureux, G.; MacKerell, J. A. D.; Roux, B. A Simple Polarizable Model of Water based on Classical Drude Oscillators; J. Chem. Phys. **2003**, *119* (10), 5185–5197.

[67] Rick, S. W.; Stuart, S. J. Potentials and Algorithms for Incorporating Polarizability in Computer Simulations In. *Reviews in Computational Chemistry*; John Wiley & Sons, Inc: 2003; Vol. 18, pp 89–146.

[68] Masia, M. Ab Initio Based Polarizable Force Field Parametrization; J. Chem. Phys. **2008**, *128* (18), 184107.

[69] Giese, T. J.; York, D. M. Many-Body Force Field Models Based Solely on Pairwise Coulomb Screening do not Simultaneously Reproduce Correct Gas-phase and Condensed-phase Polarizability Limits; J. Chem. Phys. **2004**, *120* (21), 9903–9906.

[70] Warren, G. L.; Patel, S. Electrostatic Properties of Aqueous Salt Solution Interfaces: A Comparison of Polarizable and Nonpolarizable Ion Models; J. Phys. Chem. B **2008**, *112* (37), 11679–11693.

[71] Wick, C. D.; Xantheas, S. S. Computational Investigation of the First Solvation Shell Structure of Interfacial and Bulk Aqueous Chloride and Iodide Ions; J. Phys. Chem. B **2009**, *113* (13), 4141–4146.

[72] Rogers, D. M.; Beck, T. L. Quasichemical and Structural Analysis of Polarizable Anion Hydration; J. Chem. Phys. **2010**, *132* (1), 014505.

[73] Lee, A. J.; Rick, S. W. The Effects of Charge Transfer on the Properties of Liquid Water; J. Chem. Phys. **2011**, *134* (18), 184507.

[74] Ben-Amotz, D. Unveiling Electron Promiscuity; J. Phys. Chem. Lett. **2011**, *2* (10), 1216–1222.

[75] Landau, L. D.; Lifshitz, E. M.; Pitaevski, L. P. *Electrodynamics of Continuous Media*; Butterworth-Heinemann: Pergamon, New York, 1984.

[76] Trasatti, S. The Absolute Electrode Potential: An Explanatory Note; Pure Appl. Chem. **1986**, *58* (7), 955–966.

[77] Pratt, L. R. Contact Potentials of Solution Interfaces: Phase Equilibrium and Interfacial Electric Fields; J. Phys. Chem. **1992**, *96* (1), 25–33.

[78] Parsons, D. F.; Ninham, B. W. Importance of Accurate Dynamic Polarizabilities for the Ionic Dispersion Interactions of Alkali Halides; Langmuir **2010**, *26* (3), 1816–1823.

[79] Parsegian, V. A. *Van der Waals Forces: A Handbook for Biologists, Chemists, Engineers, and Physicists*; Cambridge University Press: New York, 2006.

[80] Ninham, B. W.; Nostro, P. L. Molecular Forces and Self Assembly In *Colloid, Nano Sciences and Biology*; Cambridge University Press: New York, 2010.

[81] Dzyaloshinskii, I. E.; Lifschitz, E. M.; Pitaevskii, L. P. General Theory of van der Waals Forces; Soviet Phys. Uspekhi **1961**, *73*, 153–176 translated by Hamermesh, M. from *Usp. Fiz. Nauk*. 73, 1961 381–422.

[82] Parsons, D. F.; Deniz, V.; Ninham, B. W. Nonelectrostatic Interactions Between Ions with Anisotropic Ab Initio Dynamic Polarisabilities; Colloids Surf. A Physicochem. Eng. Asp. **2009**, *343* (13), 57–63.

[83] van der Hoef, M. A.; Madden, P. A. Three-Body Dispersion Contributions to the Thermodynamic Properties and Effective Pair Interactions in Liquid Argon; J. Chem. Phys. **1999**, *111* (4), 1520–1526.

[84] Iuchi, S.; Izvekov, S.; Voth, G. A. Are Many-Body Electronic Polarization Effects Important in Liquid Water? J. Chem. Phys. **2007**, *126* (12), 124505.

[85] Maurer, P.; Laio, A.; Hugosson, H. W.; Colombo, M. C.; Rothlisberger, U. Automated Parametrization of Biomolecular Force Fields from Quantum Mechanics/Molecular Mechanics (QM/MM) Simulations through Force Matching; J. Chem. Theory Comput. **2007**, *3* (2), 628–639.

[86] Shah, J. K.; Asthagiri, D.; Pratt, L. R.; Paulaitis, M. E. Balancing Local Order and Long-Ranged Interactions in the Molecular Theory of Liquid Water; J. Chem. Phys. **2007**, *127*, 144508.

[87] Pollack, G. L. Why Gases Dissolve in Liquids; Science **1991**, *251*, 1323–1330.

[88] Chandler, D.; Weeks, J. D.; Andersen, H. C. Van der Waals Picture of Liquids, Solids, and Phase-Transformations; Science **1983**, *220* (4599), 787–794.

[89] Paliwal, A.; Asthagiri, D.; Pratt, L. R.; Ashbaugh, H. S.; Paulaitis, M. E. An Analysis of Molecular Packing and Chemical Association in Liquid Water using Quasichemical Theory; J. Chem. Phys. **2006**, *124*, 224502.

[90] Weber, V.; Asthagiri, D. Communication: Thermodynamics of Water Modeled Using Ab Initio Simulations; J. Chem. Phys. **2010**, *133*, 141101.

[91] Pratt, L. Molecular Theory of Hydrophobic Effects: "She is too Mean to have her Name Repeated"; Ann. Rev. Phys. Chem. **2002**, *53*, 409–436.

[92] Reiss, H. Scaled Particle Theory of Hard Sphere Fluids to 1976 In *Statistical Mechanics and Statistical Methods in Theory and Application*; Landman, U., Ed.; Plenum Press: London, 1977; p 99.

[93] Hummer, G.; Garde, S.; García, A. E.; Pohorille, A.; Pratt, L. R. An Information Theory Model of Hydrophobic Interactions; Proc. Natl. Acad. Sci. USA **1996**, *93* (17), 8951–8955.

[94] Garde, S.; Hummer, G.; Garcia, A. E.; Paulaitis, M. E.; Pratt, L. R. Origin of Entropy Convergence in Hydrophobic Hydration and Protein Folding; Phys. Rev. Lett. **1996**, 77 (24), 4966–4968.

[95] Pratt, L. R.; Ashbaugh, H. S. Self-Consistent Molecular Field Theory for Packing in Classical Liquids; Phys. Rev. E **2003**, *68*, 021505.

[96] Ashbaugh, H. S.; Asthagiri, D.; Pratt, L.; Rempe, S. B. Hydration of Krypton and Consideration of Clathrate Models of Hydrophobic Effects from the Perspective of Quasi-Chemical Theory; Biophys. Chem. **2003**, *105*, 323–338.

[97] Ashbaugh, H. S.; Pratt, L. R. Colloquium: Scaled Particle Theory and the Length Scales of Hydrophobicity; Rev. Mod. Phys. **2006**, *78* (1), 159–178.

[98] Ashbaugh, H. S.; Pratt, L. R. Contrasting nonaqueous against aqueous solvation on the basis of scaled-particle theory; J. Phys. Chem. B **2007**, *111* (31), 9330–9336.

[99] Sabo, D.; Varma, S.; Martin, M. G.; Rempe, S. B. Studies of the Thermodynamic Properties of Hydrogen Gas in Bulk Water; J. Phys. Chem. B **2007**, *112* (3), 867–876.

[100] Pratt, L. R.; Garde, S.; Hummer, G. *Theories of Hydrophobic Effects and the Description of Free Volume in Complex Liquids.* NATO Sci. Ser. Ser. C 529 (New Approaches to Problems in Liquid State Theory); 1999 407–420.

[101] Chempath, S.; Pratt, L. R.; Paulaitis, M. E. Quasichemical Theory with a Soft Cutoff; J. Chem. Phys. **2009**, *130*, 054113.

[102] Rogers, D. M.; Beck, T. L. Modeling Molecular and Ionic Absolute Solvation Free Energies with Quasichemical Theory Bounds; J. Chem. Phys. **2008**, *129* (13), 134505.

[103] Asthagiri, D.; Ashbaugh, H. S.; Piryatinski, A.; Paulaitis, M. E.; Pratt, L. R. Non-van der Waals Treatment of the Hydrophobic Solubilities of CF_4; J. Am. Chem. Soc. **2007**, *129* (33), 10133–10140.

[104] Jiao, D.; Rempe, S. B. CO_2 Solvation Free Energy using Quasi-Chemical Theory; J. Chem. Phys. **2011**, *134* (22), 224506–224514.

[105] Lide, J. F.; Fredrikse, H. P. R. *CRC Handbook of Chemistry and Physics*, 75th ed.; CRC Press: Boca Raton, FL, 1995.

[106] Sabo, D.; Rempe, S. B.; Greathouse, J. A.; Martin, M. G. Molecular Studies of the Structural Properties of Hydrogen Gas in Bulk Water; Mol. Simul. **2006**, *32* (3), 269–278.

[107] Stell, G. Fluids with Long-Range Forces: Toward a Simple Analytic Theory In. *Statistical Mechanics Part A: Equilibrium Techniques*; Berne, B. J., Ed.; Plenum: New York, 1977; Vol. 5, pp 47–84.

[108] Weeks, J. D. Connecting Local Structure to Interface Formation: A Molecular Scale van der Waals Theory of Nonuniform Liquids; Ann. Rev. Phys. Chem. **2002**, *53*, 533–562.

[109] Hummer, G.; Pratt, L. R.; Garcia, A. E. Multistate Gaussian Model for Electrostatic Solvation Free Energies; J. Am. Chem. Soc. **1997**, *119* (36), 8523–8527.

[110] Martin, R. L.; Hay, P. J.; Pratt, L. R. Hydrolysis of Ferric Ion in Water and Conformational Equilibrium; J. Phys. Chem. A **1998**, *102*, 3565–3573.

[111] Pratt, L. R.; LaViolette, R. A. Quasi-Chemical Theories of Associated Liquids; Mol. Phys. **1998**, *94*, 909–915.

[112] Chempath, S.; Pratt, L. R.; Paulaitis, M. E. Distributions of Extreme Contributions to Binding Energies of Molecules in Liquids; Chem. Phys. Lett. **2010**, *487*, 24–27.

[113] Rogers, D. M.; Rempe, S. B. Probing the Thermodynamics of Competitive ion Binding Using Minimum Energy Structures; J. Phys. Chem. B **2011**, *115* (29), 9116–9129.

[114] Varma, S.; Rempe, S. B. Tuning Ion Coordination Architectures to Enable Selective Partitioning; Biophys. J. **2007**, *93*, 1093–1099.

[115] Bostick, D. L.; C. L. B, , III.. Selectivity in K^+ Channels is due to Topological Control of the Permeant ion's Coordinated State; Proc. Nat. Acad. Sci. USA **2007**, *104* (22), 9260–9265.

[116] Bostick, D. L.; Arora, K.; C. L. B, , III.. K^+/Na^+ Selectivity in Toy Cation Binding Site Models is determined by the 'host'; Biophys. J. **2009**, *96* (10), 3887–3896.

[117] Dobler, M. *Ionophores and Their Structures*; John Wiley and Sons: New York, 1981.

[118] Zhou, Y.; Morais-Cabral, J. H.; Kaufman, A.; MacKinnon, R. Chemistry of Ion Coordination and Hydration Revealed by a K^+ Channel-Fab Complex at 2.0 Å Resolution; Nature **2001**, *414* (6859), 43–48.

[119] Varma, S.; Sabo, D.; Rempe, S. B. K^+/Na^+ Selectivity in K Channels and Valinomycin: Over-Coordination *versus* Cavity-Size Constraints; J. Mol. Biol. **2008**, *376*, 13–22.

[120] Jordan, P. C. New and Notable: Tuning a Potassium Channel – the Caress of the Surroundings; Biophys. J. **2007**, *93* (4), 1091–1092.

[121] Varma, S.; Rempe, S. B. Structural Transitions in Ion Coordination Driven by Changes in Competition for Ligand Binding; J. Am. Chem. Soc. **2008**, *130* (46), 15405–15419.

[121a] Rogers, D. M.; Rempe, S. B. Reply to "Comment on 'Probing the Thermodynamics of Competitive Ion Binding Using Minimum Energy Structure'"; J. Phys. Chem. B **2012**, *116* (27) 7994–7995.

[122] McQuarrie, D. A. *Statistical Mechanics*; Harper Collins: New York, 1973.

[123] Chandler, D.; Pratt, L. R. Statistical Mechanics of Chemical Equilibria and Intramolecular Structures of Nonrigid Molecules in Condensed Phases; Journal of Chemical Physics **1976**, *65*, 2925–2940.

[124] Leung, K.; Rempe, S. B.; von Lilienfeld, O. A. Ab Initio Molecular Dynamics Calculations of Ion Hydration Free Energies; J. Chem. Phys. **2009**, *130*, 204507–204517.

[125] Rempe, S. B.; Leung, K. Response to "Comment on 'Ab Initio Molecular Dynamics Calculation of Ion Hydration Free Energies' [J. Chem. Phys. 133, 047103 (2010)]"; The Journal of Chemical Physics **2010**, *133* (4), 047104.

[126] Alam, T. M.; Hart, D.; Rempe, S. L. B. Computing the ^{7}Li NMR Chemical Shielding of Hydrated Li^+ Using Cluster Calculations and Time-Averaged Configurations from Ab Initio Molecular Dynamics Simulations; Phys. Chem. Chem. Phys. **2011**, *13* (30), 13629–13637.

[127] Varma, S.; Rempe, S. B. Coordination Numbers of Alkali Metal Ions in Aqueous Solutions; Biophys. Chem. **2006**, *124*, 192–199.

[128] Rempe, S. B.; Pratt, L. R.; Hummer, G.; Kress, J. D.; Martin, R. L.; Redondo, A. The Hydration Number of Li^+ in Liquid Water; J. Am. Chem. Soc. **2000**, *122* (5), 966–967.

[129] Rempe, S. B.; Pratt, L. R. The Hydration Number of Na^+ in Liquid Water; Fluid Phase Equil. **2001**, *183–184*, 121–132.

[130] Asthagiri, D.; Pratt, L. R. Quasi-Chemical Study of Be^{2+} (aq) Speciation; Chem. Phys. Lett. **2003**, *371* (5), 613–619.

[131] Rempe, S. B.; Asthagiri, D.; Pratt, L. R. Inner Shell Definition and Absolute Hydration Free Energy of K^+ (aq) on the basis of Quasi-Chemical Theory and Ab Initio Molecular Dynamics; Phys. Chem. Chem. Phys. **2004**, *6* (8), 1966–1969.

[132] Jiao, D.; Leung, K.; Rempe, S. B.; Nenoff, T. M. First Principles Calculations of Atomic Nickel Redox Potentials and Dimerization Free Energies: A Study of Metal Nanoparticle Growth; J. Chem. Theor. Comput. **2011**, 7, 485–495.

[133] Lockless, S.; Zhou, M.; MacKinnon, R. Structural and Thermodynamic Properties of Selective Ion Binding in a K^+ Channel; PLoS **2007**, *5*, e121.

[134] Thompson, A.; Kim, I.; Panosian, T.; Iverson, T.; Allen, T.; Nimigean, C. Mechanism of Potassium-Channel Selectivity Revealed by Na^+ and Li^+ Binding Sites within the KcsA Pore; Nat. Struct. Mol. Biol. **2009**, *16*, 1317–1324.

[135] Shealy, R.; Murphy, A.; Ramarathnam, R.; Jakobsson, E.; Subramaniam, S. Sequence-Function Analysis of the K^+-Selective Family of Ion Channels using a Comprehensive Alignment and the KcsA Channel Structure; Biophys. J. **2003**, *84*, 2929–2942.

[136] Bichet, D.; Lin, Y.-F.; Ibarra, C.; Huang, C.; Yi, B.; Jan, Y.; Jan, L. Evolving Potassium Channels by Means of Yeast Selection Reveals Structural Elements Important for Selectivity; Proc. Natl. Acad. Sci. USA **2004**, *101*, 4441–4446.

[137] Bichet, D.; Grabe, M.; Jan, Y.; Jan, L. Electrostatic Interactions in the Channel Cavity as an Important Determinant of Potassium Channel Selectivity; Proc. Natl. Acad. Sci. USA **2006**, *103*, 14355–14360.

[138] Cordero-Morales, J.; Cuello, L.; Zhao, Y.; Jogini, V.; Cortes, D.; Roux, B.; Perozo, E. Molecular Determinants of Gating at the Potassium-Channel Selectivity Filter; Nat. Struct. Mol. Biol. **2006**, *13*, 311–318.

[139] Cheng, W.; McCoy, J.; Thompson, A.; Nichols, C.; Nimigean, C. Mechanism for Selectivity-Inactivation Coupling in KcsA Potassium Channels; Proc. Natl. Acad. Sci. USA **2011**, *108*, 5272–5277.

[140] Alam, A.; Jiang, Y. Structural Analysis of Ion Selectivity in the NaK Channel; Nat. Struct. Mol. Biol. **2009**, *16*, 35–41.

[141] Roux, B.; Yu, H. Assessing the Accuracy of Approximate Treatments of Ion Hydration Based on Primitive Quasichemical Theory; J. Chem. Phys. **2010**, *132* (23), 234101.

[142] Ayala, P. Y.; Schlegel, H. B. Identification and Treatment of Internal Rotation in Normal Mode Vibrational Analysis; J. Chem. Phys. **1998**, *108*, 2314.

[143] Barone, V. Vibrational Zero-Point Energies and Thermodynamic Functions Beyond the Harmonic Approximation; J. Chem. Phys. **2004**, *120* (7), 3059–3065.

[144] Glaesemann, K. R.; Fried, L. E. A Path Integral Approach to Molecular Thermo-chemistry; J. Chem. Phys. **2003**, *118* (4), 1596–1603.

[145] Miscione, G.; Stenta, M.; Spinelli, D.; Anders, E.; Bottoni, A. New Computational Evidence for the Catalytic Mechanism of Carbonic Anhydrase; Theor. Chem. Acc. Theor. Comput. Model. **2007**, *118*, 193–201.

[146] Jiao, D.; Rempe, S. B. Combined DFT and continuum calculations of pK_a in carbonic anhydrase; Biochemistry **2012**, *51* (30), 5979–5989.

[147] Andersen, O. S. Perspectives on: Ion Selectivity; J. Gen. Phys. **2011**, *137* (5), 393–395.

[148] Rogers, D. M.; Beck, T. L.; Rempe, S. B. An Information Theory Approach to Nonlinear, Nonequilibrium Thermodynamics; J. Stat. Phys. **2011**, *145* (2), 385–409.

[149] Landau, L. D.; Lifshitz, E. M.. *Course in Theoretical Physics*; Pergamon: New York, 1980; Vol. 5; § 92.

[150] Latimer, W. M.; Pitzer, K. S.; Slansky, C. M. The Free Energy of Gaseous Ions, and the Absolute Potential of the Normal Calomel Electrode; J. Chem. Phys. **1939**, 7, 108–111.

[151] McMillan, W. G., Jr.; Mayer, J. E. The Statistical Thermodynamics of Multicom-ponent Systems; J. Chem. Phys. **1945**, *13*, 276–305.

[152] Hill, T. L. *Statistical Thermodynamics*; Addison-Wesley: Reading, MA USA, 1960.

[153] Friedman, H. L.; Dale, W. D. T. Electrolyte Solutions at Equilibrium In *Statistical Mechanics Part A: Equilibrium Techniques*; Berne, B. J., Ed.; Plenum: New York, 1977; pp 85–136.

[154] Friedman, H. L. Electrolyte Solutions at Equilibrium; Ann. Rev. Phys. Chem. **1981**, *32* 1798–1204.

[155] Kusalik, P. G.; Patey, G. N. On the Molecular Theory of Aqueous Electrolyte Solutions. III. A Comparison be Tween Born–Oppenheimer and McMillan-Mayer Levels of Description; J. Chem. Phys. **1988**, *89*, 7478–7484.

[156] Ursenbach, C. P.; Dongqing, W.; Patey, G. N. Activity Coefficients of Model Aqueous Electrolyte Solutions: Sensitivity to the Short Range Part of the Interionic Potential; J. Chem. Phys. **1991**, *94*, 6782–6784.

[157] Adelman, S. A. The Effective Direct Correlation Function, an Approach to the Theory of Liquids Solutions: A New Definition of the Effective Solute Potential; Chem. Phys. Lett. **1976**, *38*, 567–570.

[158] Andersen, H. C. Cluster Methods in Equilibrium Statistics of Fluids In *Statistical Mechanics Part A: Equilibrium Techniques*; Berne, B. J., Ed.; Plenum: New York, 1977; Vol. 5, pp 1–45.

[159] Hirata, F. *Molecular Theory of Solvation*; Kluwer: Dordrecht, 2003.

[160] *Ionic Soft Matter: Modern Trends in Theory and Applications*; Henderson, D., Holovko, M., Trokhymchuk, A., Eds.; Springer: Dordrecht, The Netherlands, 2005.

[161] Fowler, R.; Guggenheim, E. A. *Statistical Thermodynamics*; Cambridge University Press: Cambridge, 1960.

[162] Onsager, L. Theories of Concentrated Electrolytes; Chem. Rev. **1933**, *13*, 73–89.

[163] Chen, Y.; Weeks, J. Local Molecular Field Theory for Effective Attractions Between like Charged Objects in Systems with Strong Coulomb Interactions; Proc. Natl. Acad. Sci. USA **2006**, *103*, 7560–7565.

[164] Rodgers, J. M.; Weeks, J. D. Local Molecular Field Theory for the Treatment of Electrostatics; J. Phys. Condens. Matter **2008**, *20* (49), 494206.

[165] Denesyuk, N. A.; Weeks, J. D. A New Approach for Efficient Simulation of Coulomb Interactions in Ionic Fluids; J. Chem. Phys. **2008**, *128* (12), 124109.

[166] Lado, F. Integral-Equation Approach to the Calculation of the Potential Distribution in a Fluid; Phys. Rev. A **1990**, *42*, 7282–7288.

[167] Widom, B. Potential-Distribution Theory and the Statistical Mechanics of Fluids; J. Phys. Chem. **1982**, *86*, 869–872.

[168] Jackson, J. L.; Klein, L. S. Potential Distribution Method in Equilibrium Statistical Mechanics; Phys. Fluids **1964**, 7, 228–231.

[169] Pratt, L. R.; Chandler, D. Effects of Solute–Solvent Attractive Forces on Hydrophobic Correlations; J. Chem. Phys. **1980**, *73*, 3434–3441.

[170] Asthagiri, D.; Merchant, S.; Pratt, L. R. Role of Attractive Methane–Water Interactions in the Potential of Mean Force Between Methane Molecules in Water; J. Chem. Phys. **2008**, *128*, 244512.

[171] Morita, T. Virial Expansion Formulae for the Microfield and Micropotential Distribution Functions and their Applications to a High Temperature Plasma; Prog. Theor. Phys. **1960**, *23*, 1211–1213.

[172] Jackson, J. L.; Klein, L. S. Continuum Theory of a Plasma; Phys. Fluids **1964**, 7, 232–241.

[173] Oppenheim, I., Tech. Rep. ZPh-106, Convair: San Diego, 1961.

[174] Fennell, C. J.; Bizjak, A.; Vlachy, V.; Dill, K. A. Ion Pairing in Molecular Simulations of Aqueous Alkali Halide Solutions; J. Phys. Chem. B **2009**, *113* (19), 6782–6791.

[175] Fennell, C. J.; Dill, K. A. Physical Modeling of Aqueous Solvation; J. Stat. Phys. **2011**, *145* (2), 209–226.

[176] Abraham, R. J.; Cooper, M. A. The Effect of Solvent on Rotational Isomeric Equilibria; Chem. Commun. **1966**, *17*, 588–589.

[177] Marenich, A. V.; Cramer, C. J.; Truhlar, D. G. Universal Solvation Model based on Solute Electron Density and on a Continuum Model of the Solvent Defined by the Bulk Dielectric Constant and Atomic Surface Tensions; J. Phys. Chem. B **2009**, *113* (18), 6378–6396.

[178] Tomasi, J.; Mennucci, B.; Cammi, R. Quantum Mechanical Continuum Solvation Models; Chem. Rev. **2005**, *105* (8), 2999–3094.

[179] Paliwal, A.; Asthagiri, D.; Abras, D.; Lenhoff, A.; Paulaitis, M. Light-Scattering Studies of Protein Solutions: Role of Hydration in Weak Protein–Protein Interactions; Biophys. J. **2005**, *89* (3), 1564–1573.

[180] Remsing, R. C.; Rodgers, J. M.; Weeks, J. D. Deconstructing Classical Water Models at Interfaces and in Bulk; J. Stat. Phys. **2011**, *145* (2), 313–334.

[181] Beck, T. L. Hydration Free Energies by Energetic Partitioning of the Potential Distribution Theorem; J. Stat. Phys. **2011**, *145* (2), 335–354.

[182] Cacace, M. G.; Landau, E. M.; Ramsden, J. J. The Hofmeister Series: Salt and Solvent Effects on Interfacial Phenomena; Quart. Rev. Biophys. **1997**, *30* (3), 241–277.

A Review of Physics-Based Coarse-Grained Potentials for the Simulations of Protein Structure and Dynamics

Hujun Shen[*,‡] Zhen Xia[†,‡] Guohui Li[*,1] and Pengyu Ren[†,1]

[*]Molecular Modeling and Design, State Key Laboratory of Molecular Reaction Dynamics, Dalian Institute of Chemical Physics, Chinese Academy of Sciences, Dalian, P.R. China
[†]Department of Biomedical Engineering, The University of Texas at Austin, Texas, USA
[‡]Equal contribution
[1]Corresponding authors: E-mails: ghli@dicp.ac.cn, pren@mail.utexas.edu

Contents

Abstract

By simplifying the atomistic representation of a biomolecular system, coarse-grained (CG) approach enables the molecular dynamics simulation to reveal the biological processes, which occur on the time and length scales that are inaccessible to the all-atom models. Many CG physical models have been developed over the years. Here, we review the general CG force field models, which have been developed by following the fundamental physical principles. Such physics-based CG potentials have higher degree of transferability and broader applications when compared with the effective potentials, which are derived from specific molecular systems and environments. We expect growing interests in developing and applying general CG force fields at different levels of problems such as protein dynamics and structure prediction.

Annual Reports in Computational Chemistry, Volume 8
ISSN 1574-1400,
http://dx.doi.org/10.1016/B978-0-444-59440-2.00005-3

1. INTRODUCTION

Since 1977, it has been widely accepted by scientists [1] that a folded protein is an ensemble of numerous conformations fluctuating in the neighborhood of its native state. A folded protein molecule does not have a single static conformation. Protein motions play a pivotal role in biological processes [2]. Hence, the study of protein dynamics has received quite a lot of attention. Nuclear magnetic resonance spectroscopy helps us in understanding and exploring protein dynamics. However, its application is limited by protein sizes. With an increase in the computational power, atomistic molecular dynamics (MD) simulation [3] becomes attractive because the atomistic motions of a biomolecule can be elucidated as a function of time [4]. The first MD simulation of a small protein was performed in vacuum with the length of 9.5 ps in 1977. The length of the atomistic MD simulation of a typical biomolecular system has been extended to several hundred nanoseconds and microseconds today. Despite many advancements in the past decade, atomistic MD simulations, limited with a small value of the integration time step (of the order of femtoseconds), fail to correctly elucidate many biological processes associated with the large conformation change, such as the protein folding and protein–protein docking. Most of these processes occur on the time scales ranging from microseconds to seconds. It is quite difficult to examine very large bimolecular complexes made of several proteins and/or poly nucleic acids. To extend the applicability of MD simulations and study the biological processes occurring on longer time scales and larger length scales, we need to use a simplified or coarse-grained (CG) representation of a biomolecular system, instead of an atomically detailed description. A variety of CG models have been developed to study biomolecular systems. However, they need to strike a balance between accuracy and efficiency [5] (see Fig. 5.1).

Coarse-graining approach of the biomolecular system is not unique. It depends on the developers' purposes or the biological phenomena of interest [6]. Numerous CG models have been devised. The elastic network model (ENM) [7–9] is based on the seminal work of Flory et al. [10] on polymer networks and exploration of protein dynamics. It uses a normal mode analysis and simplified harmonic potentials [11] and has been proved to be useful in analyzing the slow motions of a protein. ENM can be used to calculate vibrational entropy efficiently and accurately [12]. Moreover, ENM can be combined with MARTINI CG force field to study protein dynamics and interactions between protein and lipid bilayer [13,14], which

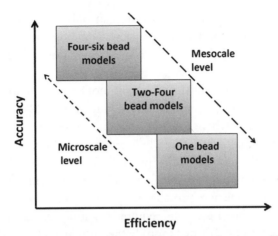

Figure 5.1 Schematic illustration of CG models for biomolecules at different levels.

is termed as ELNEDIN [15]. In the ELNEDIN approach, the secondary and tertiary structures of a protein are modeled in the form of ENM. A harmonic restraint with a force constant is applied to all the backbone particles within a distance of each other. MARTINI CG force field was originally developed for modeling lipids and surfactants [16,17]. It has been extended to simulate proteins [18] and carbohydrates [19]. In recent times, a hybrid scheme, combining MARTINI CG with the atomistic model, was proposed after taking into consideration the accuracy of an atomistic model and the sampling speed of the CG model [20]. Nevertheless, both ENM and MARTINI models are only suitable for near native dynamics. They fail to explore the protein folding or predict structures.

Go's models have been quite useful in studying the protein-folding pathway, as they employ structure-based approaches [21]. In the native state, a protein structure is composed of backbone atoms with secondary structures (helix, β-strand or coil) and highly well-packed side chains in its core. The non-native protein structure is described with low-level secondary structure packing, which is associated with flexible side chains. Due to the low level of side chain packing in the non-native state but high-level packing in the native state, the CG model can describe a large fluctuation of an unfolded protein. But, if the side chain information (e.g. shape and specific interaction) is largely lost in coarse graining, it becomes difficult to capture the native structure of proteins.

In Go models, energy terms are constructed in favor of the native contacts, whereas the non-native contacts are termed as less favorable or repellent. In fact, non–native interactions can only contribute to the local structural perturbations or stabilize the protein-folding transition states along the protein-folding pathway. Its transition states are primarily determined by native interactions [22].

Protein dynamics near the native state cannot be characterized completely by the original on-lattice Go model [23]. Onuchic et al. [24] introduced the off-lattice Go model, wherein the protein fluctuations near the native state are considered as quasi-harmonic. These approached the funnel-like energy landscape [25] of the protein-folding process. Kenzaki et al. devised CafeMol, a CG simulator adopting this off-lattice Go model, with the purpose of developing and simulating proteins [26]. Nevertheless, since the native information of a protein is usually required for structure-based CG models, these approaches cannot be used to study the protein-folding pathway. They cannot be used for predicting the structure of proteins.

Tanaka and Scheraga [28] first proposed the knowledge-based statistical potentials [27] for predicting the structure of proteins. They can be derived from the statistical distribution of native structures in the protein data bank (PDB) through Boltzmann-based methods. Miyazawa and Jernigan [29] followed up the work of Tanaka and Scheraga. They introduced the effect of solvent into the potential. Godzik and Skolnick [30] defined the statistical potentials by adding the residue triplet's term. In addition, the dihedral angle term and other statistical terms, such as solvent accessibility and hydrogen bonding have been added to the knowledge-based statistical potential [31,32]. A log-linear model was developed by Bryant and Lawrence [33]. According to this model, it is inappropriate to derive the empirical potential from simple summation over the distribution of residues in the proteins. Rather, protein structures need to be analyzed specifically and separately.

These knowledge-based approaches have been employed in simulations of protein folding and protein structure prediction. They had considerable successes [34–36] over the past decade. According to the knowledge-based approaches, interactions can be assumed to be pairwise, and the statistical distributions can be considered Boltzmann related. Thus, the pairwise interaction free energy $F_{ij}(r)$ can be calculated from the ratio of observed probability $P_{ij}(r)$ as a function of pairwise distance r_{ij} to the probability P^0 of the reference state, wherein the zero interaction is defined as follows:

$$F_{ij}(r) = -kT\ln\left(\frac{P_{ij}}{P^0}\right), \qquad (5.1)$$

where k is the Boltzmann's constant and T is the absolute temperature.

However, in recent times, knowledge-based approaches have been criticized because their fundamental basis and physical meaning are questionable [37]. First, the treatment of many-body correlation in the knowledge-based approach is approximate due to the nontrivial contribution from the packing effects [38–40]. Second, the treatment of statistical distributions from the different protein structures in PDB fails to provide an accurate structural fluctuation of a single protein at equilibrium [41,42].

Following the seminal work of Levitt and Warshel [43], a variety of physics-based CG approaches have been proposed. They have proved to be useful for both the protein-folding pathway and protein dynamics near the native state. In fact, they can also be used for predicting protein structure. Among them, one-bead CG model, developed by Tozzini and McCammon [44], has been employed successfully to study the flap opening of HIV-1 protease. More complex CG models, such as the UNRES model [45–47], have been developed to predict the protein structures. In UNRES, a polypeptide chain is composed of a sequence of α-carbons with specific united side chains and united peptide groups. These CG particles are connected through virtual bonds. To parameterize the various energy terms, both quantum mechanical ab initio methods and all-atom MD simulations were used.

Another physics-based CG model, namely the GB-EMP model, consists of rigid bodies of anisotropic Gay–Berne ellipsoidal particles and point multipoles [48–50]. Other CG models are constructed by the intermolecular interaction energy and force at the functional group or molecular level, using the all-atom simulations of specific systems. The focus of the GB-EMP model is based on representing the components of the intermolecular forces, such as the electrostatic and repulsion–dispersion at the CG level. The strategy is very similar to that of developing an empirical atomic potential energy (PE) model based on quantum mechanical principles. In addition, because of the anisotropic shape of the CG particle and point multipoles sharing the same local frame with the Gay–Berne particle, the side chain information of any amino acid can be retained. Therefore, the GB-EMP model is designed to be transferable and not limited by specific systems or environments.

In this review, we will present several representatives of physics-based models. One-bead, UNRES and GB-EMP CG models and applications of physics-based models were used to study the protein dynamics and protein structure. For other CG models, such as the energy-based MARTINI model and structure-based CNM-CG model, you may refer to the previous review by Merchant and Madura [6]. In addition, the CG potentials we discussed here can be distinguished from the *coarse-graining approaches*, such as force matching [51–57], which have been reviewed widely.

2. CG MODELS

2.1. One-Bead CG Model

In one-bead models, each amino acid of a protein is grouped into one bead centered at α-carbon C_α. A polypeptide chain is represented through a sequence of α-carbons connected by virtual bonds (Fig. 5.2). Therefore, the backbone conformation, described by the two backbone dihedral angles ϕ and ψ in an all-atom representation, is totally different from that of the CG representation, such as θ and α, where the angle θ is determined through three consecutive α-carbons and the dihedral angle α is determined through four consecutive α-carbons. The energy function is a sum of the stretching bond term U_b, a bending angle term U_θ, a dihedral angle term U_α and non-bonded term U_{nb}, such as

$$U = U_b + U_\theta + U_\alpha + U_{nb} \qquad (5.2)$$

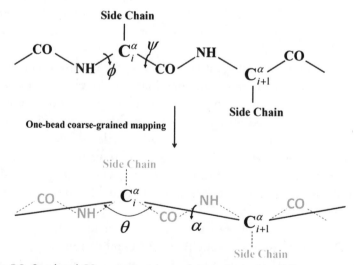

Figure 5.2 One-bead CG mapping scheme for a polypeptide chain.

The functional form of the dihedral term U_α could be adopted from the approach of Sorenson and Head-Gordon [58], such as

$$U_\alpha = \sum_{\text{dihedrals}} A[1 + \cos\alpha] + B[1 - \cos\alpha]$$
$$+C[1 + \cos 3\alpha] + D[1 + \cos(\alpha + \pi/4)]$$

(5.3)

The non-bonded term is separated into two parts: a local anisotropic term $U_{\text{nb-loc}}$ and a nonlocal isotropic term $U_{\text{nb-nonloc}}$ [59].

In principle, physics-based one-bead CG model should have a high degree of transferability, when no reference protein structures are used to parameterize its energy potentials. However, the accuracy of protein structure prediction is limited by its two intrinsic features: (1) the bending angle term U_θ and the dihedral angle term U_α are parameterized independent of secondary structures and (2) the non-bonded term U_{nb} does not take into consideration the side chain effect and hydrogen bond propensity.

2.2. UNRES CG Model

The UNRES force field was originally developed to solve the problem of protein structure prediction. It used the global minimum of PE function. Fig. 5.3 shows the UNRES CG representation of a polypeptide. In the

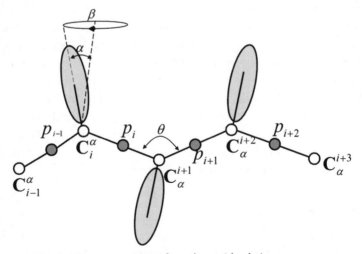

Figure 5.3 UNRES CG representation of a polypeptide chain.

UNRES model, both side chain (SC) (shaded ellipsoid) and the peptide group (p) (shaded circles) centroids are referred as the interaction sites, while the α-carbon atoms (small empty circles) do not involve in any interaction. They are only the connection points. Two consecutive α-carbons define the interaction center of the peptide group. The backbone conformation in UNRES model is described by a virtual bond angle θ and a dihedral angle γ, while the side chain conformation is defined through two rotational angles α and β. The effective energy function, termed as restricted free energy (RFE), is given as the Eqn (5.4)

$$U = w_{SC} \sum_{i<j} U_{SC_iSC_j} + w_{SCp} \sum_{i \neq j} U_{SC_ip_j} + w_{pp} \sum_{i<j-1} U_{p_ip_j}$$

$$+w_{bond} \sum_i U_{bond}(d_i) + w_b \sum_{i \neq j} U_b(\theta_i) + w_{rot} \sum_i U_{rot}(\alpha_i, \beta_i, \theta_i)$$

$$+w_{tor} \sum_i U_{tor}(\gamma_i) + w_{tord} \sum_i U_{tord}(\gamma_i, \gamma_{i+!}) + \sum_{m=3}^{6} w_{corr}^{(m)} U_{corr}^{(m)}$$

$$+w_{SS} \sum_i U_{SS}(i)$$

(5.4)

in which $U_{SC_iSC_j}$, $U_{SC_ip_j}$, and $U_{p_ip_j}$ describe the interactions between side chains (SC), between side chain and peptide, and between peptides, respectively. $U_{p_ip_j}$ can be further split into backbone vdW term $U_{p_ip_j}^{vdW}$ and backbone–electrostatic term $U_{p_ip_j}^{el}$. U_{bond}, U_b and U_{rot} denote the effective energies for virtual bond stretching, angle bending, and rotameric states of virtual side chains, respectively. U_{tor} and U_{tord} are torsional and double-torsional potentials, respectively. $U_{corr}^{(m)}, (m = 3-6)$, related to coupling between backbone–local and backbone–electrostatic interactions, are correlated with the multibody terms [60]. U_{SS} described the effective energy of disulfide bonds and w as the weight factor for each energy term. The UNRES force field should be described as RFE function instead of PE function. Recently, temperature dependence was introduced into the force field [61,62] by utilizing Kubo's cumulant expansion [63]. This change enables the UNRES model to search the native protein structure and study the protein-folding dynamics under different temperatures.

2.3. GB-EMP CG Model

Based on the general physical principles of molecular interactions, the Gay-Berne potential with the electrostatic multipole (GB-EMP) model has been developed. Here, the fundamental components of the intermolecular forces have been represented explicitly. The van der Waals interaction is described by treating the molecules as soft uniaxial ellipsoids with a generalized anisotropic Gay–Berne function. The charge distribution is represented by off-centered multipoles, including point charge, dipole, and quadrupole moments. Similar to an all-atom force field, the parameters in GB-EMP model can apply to different environments without re-parameterization. The individual energy components in the CG model, including vdW, electrostatics, solvation, and torsional energy contributions, have been designed to match with those of the all-atom force fields, in both the gas phase and solution.

The Gay–Berne anisotropic PE is based on the Gaussian overlap potential [64]. The energy between two Gay–Berne particles i and j has the form

$$U_{GB}(\hat{u}_i, \hat{u}_j, r_{ij}) = 4\varepsilon(\hat{u}_i, \hat{u}_j, \hat{r}_{ij}) \left[\left(\frac{d_w \sigma_0}{r_{ij} - \sigma(\hat{u}_i, \hat{u}_j, \hat{r}_{ij}) + d_w \sigma_0} \right)^{12} \right.$$
$$\left. - \left(\frac{d_w \sigma_0}{r_{ij} - \sigma(\hat{u}_i, \hat{u}_j, \hat{r}_{ij}) + d_w \sigma_0} \right)^6 \right] \tag{5.5}$$

The range parameter σ and the strength parameter ε for pairwise interactions are functions of the relative orientation of particles. Each uniaxial molecule is associated with a set of Gay–Berne parameters, describing its ellipsoid shape and orientation of its principal axis in the inertial frame, which is defined according to the all-atom model. The term d_w is used to control the "softness" of the potential [65]. A detailed discussion of the interaction parameters can be found in earlier publications [48–50,66].

The charge distribution for a given Gay–Berne particle can be represented as a multipole expansion, which is placed at the center or off-center of the particle:

$$M = \left[q, d_x, d_y, d_z, Q_{xx}, Q_{xy}, \dots, Q_{zz} \right] \tag{5.6}$$

where q, d and Q are charge, dipole, and quadrupole moments, respectively. This multipole expansion gives us an accurate electrostatic potential at a distance greater than the size of the particle. It is often termed the convergence sphere [67]. The interaction energy between two multipole sites can be expressed in its polytensor form [68]. GB-EMP model was first implemented in several small molecules, where benzene and methanol molecules have been chosen to represent disk-like and rod-like particles, respectively.

In the GB-EMP liquid simulations of benzene, methanol, and water, where each molecule is represented by a particle, a range of thermodynamic and structural properties was computed from CG MD simulations. These showed an agreement with experimental values within 2%. In addition, hydrogen bonds are well captured in the methanol model. GB-EMP model has been successfully applied to study polyalanine [50]. Its extension is used to study the protein folding and dynamics. Fig. 5.4a shows the example of the alanine dipeptide model. An alanine dipeptide is composed of three

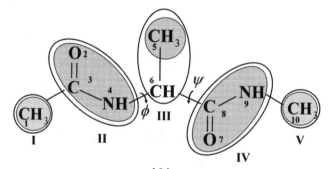

(A)

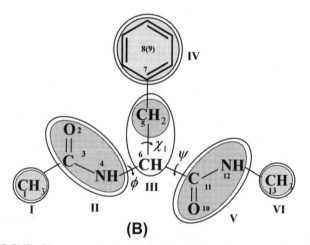

(B)

Figure 5.4 GB-EMP CG representations of (A) alanine dipeptide model and (B) phenylalanine dipeptide model. Ellipsoids encompass the rigid bodies (unshaded ellipsoids or circles) that contain Gay–Berne (shaded ellipsoids and circle) and multipole interaction sites (labeled numbers). The Gay–Berne particles are located at the centers of the mass of the corresponding atoms.

spherical methyl groups and two ellipsoid amide groups. Each group contains Gay–Berne and/or electrostatic multipole sites. Two different rigid bodies are connected through a virtual bond or covalent bond between Gay–Berne and/or electrostatic multipole sites.

3. SOLVENT MODELS

Liquid water and aqueous solutions are crucial for the structure and function of biomolecules. Yet, it is computationally expensive to describe the protein solvation with the help of atomistic models. In principle, a good CG water model greatly increases the computational efficiency and orders of magnitude in the length scale of a protein–solvent system. This describes the solvation in a reasonable manner. Similar to the atomistic models, a CG model treats the solvent effect either explicitly or implicitly. In the explicit CG solvent models, a single particle is used to represent one or a cluster of water molecules.

The "sticky dipole" potential water model was initially developed by Bratko et al. [69], and it is referred to as the BBL model. BBL model has a single site at the molecular center of the mass with a spherically repulsive potential, a short-range tetrahedral "sticky" potential, and a point dipolar potential. Later, Ichiye and Liu improved the model by introducing a Lennard-Jones soft-sphere potential instead of the original hard-sphere model in order to better fit biological molecules [70]. Recently, they introduced higher order moments of the electrical charge to accurately mimic the PE function of a multipoint model [71]. Water molecules can be more crudely CG by combining them together in a cluster of waters.

For instance, MARTINI polarizable water model treats four-water clusters as one CG bead [72]. Similarly, Wu et al. [73] proposed a four-water bead model, namely a big multiple water (BMW) model. This suggested that the computational efficiency and accuracy of the four-water bead model can be well balanced, which is supported by other investigators [74]. In the BMW model, each CG unit consists of three charged sites and an additional non–electrostatic soft interaction site at the unit center.

The polarizable pseudo-particles model is a semi-implicit water model. It was proposed to combine the advantages of both the explicit model and implicit water models [75–79]. Each water molecule (which can be extended to other solvents) has a Lennard-Jones particle site using polarizable point dipoles. The point dipoles cannot interact with each other but only with the solute. Therefore, the non-bonded interactions between the solvents are reduced to only simple and efficient short-range Lennard-Jones potentials. The advantages of such a method are that the polarizable dipoles do not have to be iterated to self-consistency because they simply respond to the field of the solute. In addition, the vdW interactions of the solvent are included in the potential, which are negligible in the implicit solvent models.

In implicit solvent models, the electrostatic energy of the charged atom is calculated using the macroscopic continuum theory. A general analytic solution to the Poisson equation is solved only through approximations of an arbitrarily spaced collection of spherical dielectric particles, which are embedded in the solvent. For example, for realistically exploring the molecular geometries and monopole charge distributions, the generalized Born (GB) approach provides an approximate numerical solution of the Poisson equation [80–92].

In recent times, for the electrostatic component of the solvation free energy, Schnieders and Ponder developed a continuum electrostatics model based on Kirkwood's analytic result. In this case, the solute electrostatics are treated by permanent and induced atomic multipole moments: these are not limited to monopoles like in GB [93]. The generalized Kirkwood (GK) method has been successfully applied to the newly developed atomic multipole optimized energetics for biomolecular applications (AMOEBA) force field [94,95]. Later, it was applied to the generalized CG model based on Gay–Berne and electrostatic multipole for peptide systems [50]. With the GK approach, the computational efficiency was improved significantly for the physically based CG model. As for the one-bead and UNRES CG models, the implicit solvent effect is taken into consideration in the parameterization of $U_{nb-nonloc}$ [59] and $U_{SC_iSC_j}$ [96], respectively.

4. PROTEIN MODELING
4.1. Peptide Backbone Conformation

To evaluate the conformational properties of a CG model, the conformational energy map of alanine dipeptide was used as a function of backbone dihedral angles (ϕ and ψ). In general, backbone conformation of a dipeptide CG model is defined through bending angles and dihedral angles. For instance, as we mentioned earlier, the backbone conformation of the one-bead CG model is defined by θ, α, while that of the UNRES CG model is described by θ, γ. In the UNRES CG model, the center of the peptide group is placed at the geometric center of two consecutive α-carbons. So, the dihedral angle γ defined in the UNRES model is equivalent to the dihedral angle α in one-bead CG model. In the GB-EMP CG model, the atomistic description of the backbone conformation (ϕ and ψ) can be retained. In general, by comparing the conformation energy maps obtained from CG models with the corresponding energies of atomistic models, one can find that the CG energy surfaces are smoother than that of atomistic models. The smooth nature of coarse graining is attributed to a reduction in the degrees of freedom. This in turn reduces the number of atomistic interactions.

While comparing the results from the simulation of the 12-residue polyalanine, the dihedral angles ϕ and ψ distribution for 5-residue poly-alanine are found to be significantly different. In this case, the population of β-strand obviously increases, while that of α-helix population changes slightly [50]. Note that earlier all-atom simulation calculations [97,98] have shown the same phenomena. Best et al. [97] used CHARMM27/cmap force field to sample the conformational space of the 5-mer polyalanine system. They reported that 57.5% and 19.8% are produced for α-helix and β-populations, respectively, in the ϕ and ψ dihedral angle distribution. The OPLSAA/L force field yielded a little bit lower population for α-helix and surprisingly the same population for β-strand. Likewise, the population of full α-helices, in which all the five alanine residues adopt α-helical conformation in the ϕ and ψ dihedral angle distribution, is given at 4.62% when compared to 8% and 1% produced in the atomistic simulations of CHARMM and AMBER03 force fields, respectively [98]. Although the statistical distribution of ϕ and ψ dihedral angle, obtained from circular dichroism spectroscopy and Fourier-transform infrared experiments, sug-gested that the helical content in the 5-mer has been overestimated in atomistic simulations using Amber03, CHARMM27/cmap, and OPLA-aa/L force field [98]. Physics-based CG models seem to qualitatively reproduce the distributions sampled from the atomistic MD simulations, which is not surprising, since the atomistic models are often used to parameterize CG models. Statistical population from PDB can be alternatively used to parameterize the CG conformational properties.

4.2. Protein Folding and Dynamics

The simplest CG models of proteins, such as the one-bead CG model, are typically constructed at a highly coarse level. The information about the interactions between side chain–side chain and side chain–backbone play an important role in protein folding.

To attack the protein-folding problem, both physics-based CG and the atomistic models share the same underlying assumption, that is, the Anfinsen's thermodynamic hypothesis [99]. The native structure of the protein, under certain conditions, should have a minimal free energy of the protein. Thus, the free energy on the folding pathway should be associated with the order of the forming native-like structure of the protein. This assumption forms the fundamental basis of developing UNRES CG force field for protein folding [100]. However, search engines, such as the conformational space annealing (CSA) [101], employed in the UNRES model, favor α-helical structures over β-structures. Conversely, UNRES cannot produce correct α-helical structures when the energy function is optimized to produce β-structure proteins. Thus, the replica exchange method, in place of CSA, has become a new search engine for the UNRES model [62].

During enzyme catalysis, a protein may experience a large conformation change, which can be successfully detected by one-bead physic-based CG models. For instance, the one-bead CG model, developed by Tozzini and McCammon [44], is applied to HIV-1 protease, wherein the opening frequency occurs on a microsecond–millisecond time scale. Meanwhile, this model was employed to study ligand and substrate binding to HIV protease, as well as the ligand binding pathway [102–104].

The success of one-bead models is attributed to the following reasons: (1) one-bead models are the coarsest models among all the current physics-based CG models, which are computationally efficient and still retain the main dynamic features of a protein; (2) many important biological processes involve the transition in secondary structures as the rate-limited step, which can be described well by one-bead models; and (3) many biological functions are associated with slow motions in which all the details of protein structures are not required to be known. However, in some cases, orientation or rotation mode of side chains plays a critical role in protein–protein or protein–ligand docking. In these cases, higher resolution CG models, such as UNRES, GB-EMP models, and other four–six-bead models or higher level CG models are required.

5. CONCLUDING REMARKS

Coarse graining is able to correctly represent some properties of interest on the time and length scales, which are inaccessible to atomistic models. Coarse gaining reduces the number of degrees of freedom of a system while retaining its major physical features. Due to the reduction in the number of particles and larger simulation time steps, the computational efficiency is generally enhanced by two to three orders of magnitude when compared with the all-atom simulations.

The coarse-graining strategy is not unique, and it can be conducted at different detail levels, accuracy, and efficiency. When all the dynamic details of a system are not required to be captured, low-resolution CG models, such as one-bead models, are superior to high-resolution CG models, such as UNRES, GB-EMP, and other four–six-bead models. However, high-resolution CG models are required when some dynamic details associated with local conformation changes, such as side chain rotation or side chain packing, need to be known.

The accuracy and transferability of CG potentials depends on the choice of energy functions and parameterization process. In this review, we presented a few CG models at different CG levels. They are coined physics-based CG models due to the physical consideration taken in their parameterizations of effective energy functions. This is in contrast to the "statistical potential" approach, where the atomic forces in specific molecular environments are lumped into effective potential. In general, the physics-based CG models share the same development philosophy as the atomistic molecular mechanics model. They follow the fundamental principles of intermolecular forces, which could be more transferable. Parameterization of such models is quite important and can be more challenging, given the uncertainty of the atomistic potentials we rely on.

REFERENCES

[1] McCammon, J. A.; Gelin, B. R.; Karplus, M. Dynamics of Folded Proteins; *Nature* **1977**, *267*, 585–590.

[2] Karplus, M.; Kuriyan, J. Molecular Dynamics and Protein Function; *Proc. Natl. Acad. Sci. U.S.A.* **2005**, *102*, 6679–6685.

[3] Adcock, S. A.; McCammon, J. A. Molecular Dynamics: A Survey of Methods for Simulating the Activity of Proteins; *Chem. Rev.* **2006**, *106*, 1589–1615.

[4] Karplus, M.; McCammon, J. A. Molecular Dynamics Simulations of Biomolecules; *Nat. Struct. Biol.* **2002**, *9*, 646–652.

[5] Tozzini, V. Coarse-Grained Models for Proteins; *Curr. Opin. Struct. Biol.* **2005**, *15*, 144–150.

[6] Merchant, B. A.; Madura, J. D. A Review of Coarse-Grained Molecular Dynamics Techniques to Access Extended Spatial and Temporal Scales in Biomolecular Simulations; *Annu. Rep. Comput. Chem.* **2011**, *7*, 67–85.

[7] Tirion, M. M. Large Amplitude Elastic Motions in Proteins from a Single-Parameter, Atomic Analysis; *Phys. Rev. Lett.* **1996**, *77*, 1905–1908.

[8] Bahar, I.; Atilgan, R.; Erman, B. Direct Evaluation of Thermal Fluctuations in Protein using a Single Parameter Harmonic Potential; *Fold. Des.* **1997**, *2*, 173–181.

[9] Haliloglu, T.; Bahar, I.; Erman, B. Gaussian Dynamics of Folded Proteins; *Phys. Rev. Lett.* **1997**, *79*, 3090–3093.

[10] Flory, P. J.; Gordon, M.; McCrum, N. G. Statistical Thermodynamics of Random Networks; *Proc. Roy. Soc. Lond. A* **1976**, *351*, 351–380.

[11] Go, N.; Noguti, T.; Nishikawa, T. Dynamics of a Small Globular Protein in Terms of Low-Frequency Vibrational Modes; *Proc. Natl. Acad. Sci. U.S.A.* **1983**, *80*, 3696–3700.

[12] Xu, B., et al. Fast and Accurate Computation Schemes for Evaluating Vibrational Entropy of Proteins; *J. Comput. Chem.* **2011**, *32*, 3188–3193.

[13] Balali-Mood, K.; Bond, P. J.; Sansom, M. S. P. Interaction of Monotopic Membrane Enzymes with a Lipid Bilayer: A Coarse-Grained MD Simulation Study; *Biochemistry* **2009**, *48*, 2135–2145.

[14] Bond, P. J., et al. Assembly of Lipoprotein Particles Revealed by Coarse-Grained Molecular Dynamics Simulations; *J. Struct. Biol.* **2007**, *157*, 579–592.

[15] Periole, X., et al. Combining an Elastic Network with a Coarse-Grained Molecular Force Field: Structure, Dynamics and Intermolecular Recognition; *J. Chem. Theory Comput.* **2009**, *5*, 2531–2543.

[16] Marrink, S. J.; de Vries, A. H.; Mark, A. E. Coarse Grained Model for Semi-quantitative Lipid Simulations; *J. Phys. Chem. B* **2004**, *108*, 750–760.

[17] Marrink, S. J., et al. The MARTINI Forcefield: Coarse Grained Model for Biomolecular Simulations; *J. Phys. Chem. B* **2007**, *111*, 7812–7824.

[18] Monticelli, L., et al. The MARTINI Coarse-Grained Force Field: Extension to Proteins; *J. Chem. Theory Comput.* **2008**, *4*, 819–834.

[19] Lopez, C. A., et al. The Martini Coarse Grained Force Field: Extension to Carbohydrates; *J. Chem. Theory Comput.* **2009**, *5*, 3195–3210.

[20] Rzepiela, A. J., et al. Hybrid Simulations: Combining Atomistic and Coarse-Grained Force Fields using Virtual Sites; *Phys. Chem. Chem. Phys.* **2011**, *13*, 10437–10448.

[21] Go, N.; Taketomi, H. Respective Roles of Short-Range and Longrange Interactions in Protein Folding; *Proc. Natl. Acad. Sci. U.S.A.* **1978**, *75* (2), 559–563.

[22] Karanicolas, J.; Brooks, C. L. Improved Go -like Models Demonstrate the Robustness of Protein Folding Mechanisms Towards Nonnative Interactions; *J. Mol. Biol.* **2003**, *334*, 309–325.

[23] Go, N. Theoretical Studies of Protein Folding; *Annu. Rev. Biophys. Bioeng.* **1983**, *12*, 183–210.

[24] Clementi, C.; Nymeyer, H.; Onuchic, J. Topological and Energetic Factors: What Determines the Structural Details of the Transition State Ensemble and "en-route" Intermediates for Protein Folding? An Investigation for Small Globular Proteins; *J. Mol. Biol.* **2000**, *298*, 937–953.

[25] Leopold, P.; Montal, M.; Onuchic, J. Protein Folding Funnels: A Kinetic Approach to the Sequence-Structure Relationship; *Proc. Natl. Acad. Sci. U.S.A.* **1992**, *89*, 8721–8725.

[26] Kenzaki, H., et al. CafeMol: A Coarse-Grained Biomolecular Simulator for Simulating Proteins at Work; *J. Chem. Theory Comput.* **2011**, *7*, 1979–1989.

[27] Sippl, M. J. Knowledge-Based Potentials for Proteins; *Curr. Opin. Struct. Biol.* **1995**, *5*, 229–235.

[28] Tanaka, S.; Scheraga, H. A. Medium- and Long-Range Interaction Parameters between Amino Acids for Predicting Three-Dimensional Structures of Proteins; *Macromolecules* **1976**, *9*, 945–950.

[29] Miyazawa, S.; Jernigan, R. L. Estimation of Effective Interresidue Contact Energies from Protein Crystal Structures: Quasi-Chemical Approximation; *Macromolecules* **1985**, *18*, 534–552.

[30] Godzik, A.; Skolnick, J. Sequence-Structure Matching in Globular Proteins: Application to Supersecondary and Tertiary Structure Determination; *Proc. Natl. Acad. Sci. U.S.A.* **1992**, *89*, 12098–12102.

[31] Kocher, J. P.; Rooman, M. J.; Wodak, S. J. Factors Influencing the Ability of Knowledge-Based Potentials to Identify Native Sequence-Structure Matches; *J. Mol. Biol.* **1994**, *16*, 1598–1613.

[32] Nishikawa, K.; Matsuo, Y. Development of Pseudoenergy Potentials for Assessing Protein 3-D-1-D Compatibility and Detecting Weak Homologies; *Protein Eng.* **1993**, *6*, 811–820.

[33] Bryant, S. H.; Lawrence, C. E. An Empirical Energy Function for Threading Protein Sequence through the Folding Motif; *Proteins: Struct. Funct. Genet.* **1993**, *16*, 92–112.

[34] Buchete, N. V.; S. J, E.; Thirumalai, D. Development of Novel Statistical Potentials for Protein Fold Recognition; *Curr. Opin. Struct. Biol.* **2004**, *14*, 225–232.

[35] Lazaridis, T.; Karplus, M. Effective Energy functions for Protein Structure Prediction; *Curr. Opin. Struct. Biol.* **2000**, *10*, 139–145.

[36] Skolnick, J. In Quest of an Empirical Potential for Protein Structure Prediction; *Curr. Opin. Struct. Biol.* **2006**, *16*, 166–171.

[37] Moult, J. Comparison of Database Potentials and Molecular Mechanics Forcefields; *Curr. Opin. Struct. Biol.* **1997**, *7*, 194–199.

[38] Betancourt, M. R. Another Look at the Conditions for the Extraction of Protein Knowledge-Based Potentials; *Proteins* **2009**, *76*, 72–85.

[39] Dehouck, Y.; Gilis, D.; Rooman, M. J. A new generation of statistical potentials for proteins; *Biophys. J.* **2006**, *90*, 4010–4017.

[40] Skolnick, J., et al. Derivation and Testing of Pair Potentials for Protein Folding. When is the Quasichemical Approximation Correct? *Protein Sci.* **1997**, *6*, 676–688.

[41] Ben-Naim, A. Statistical Potentials Extracted from Protein Structures: Are These Meaningful Potentials? *J. Chem. Phys.* **1997**, *107*, 3698–3706.

[42] Thomas, P. D.; Dill, K. A. Statistical Potentials Extracted from Protein Structures: How Accurate are They? *J. Mol. Biol.* **1996**, *257*, 457–469.

[43] Levitt, M.; Warshel, A. Computer Simulation of Protein Folding; *Nature* **1975**, *235*, 694–698.

[44] Tozzini, V.; McCammon, J. A. A Coarse-Grained Model for the Dynamics of Flap Opening in HIV-1 Protease; *Chem. Phys. Lett.* **2005**, *413*, 123–128.

[45] Liwo, A., et al. Calculation of Protein Backbone Geometry from Alpha-Carbon Coordinates Based on Peptide-Group Dipole Alignment; *Protein Sci.* **1993**, *2*, 1697–1714.

[46] Liwo, A., et al. Prediction of Protein Conformation on the Basis of a Search for Compact Structures: Test on Avian Pancreatic Polypeptide; *Protein Sci.* **1993**, *2*, 1715–1731.

[47] Ołdziej, S., et al. Physics-Based Protein-Structure Prediction using a Hierarchical Protocol Based on the UNRES Force Field: Assessment in Two Blind Tests; *Proc. Natl. Acad. Sci. U.S.A.* **2005**, *102*, 7547–7552.

[48] Golubkov, P. A.; Ren. Generalized Coarse-Grained Model Based on Point Multipole and Gay–Berne Potentials; *J. Chem. Phys.* **2006**, *125*, 064103.

[49] Golubkov, P. A.; Wu, J. C.; Ren. A Transferable Coarse-Grained Model for Hydrogen Bonding Liquids; *Phys. Chem. Chem. Phys.* **2008**, *10*, 2050–2057.

[50] Wu, J., et al. Gay–Berne and Electrostatic Multipole Based Coarse-Grain Potential in Implicit Solvent; *J. Chem. Phys.* **2011**, *135* (15), 155104.

[51] Izvekov, S.; Voth, G. A. A Multiscale Coarse-Graining Method for Biomolecular Systems; *J. Phys. Chem. B* **2005**, *109*, 2469–2473.

[52] Izvekov, S.; Voth, G. A. Multiscale Coarse-Graining of Liquid State Systems; *J. Chem. Phys.* **2005**, *123*, 134105.

[53] Das, A.; Andersen, H. C. The Multiscale Coarse-Graining Method. III. A Test of Parrwise Additivity of the Coarse-Grained Potential and of New Basis Functions for the Variational Calculation; *J. Chem. Phys.* **2009**, *131*, 034102.

[54] Krishna, V.; Noid, W.; Voth, G. A. The Multiscale Coarse-Graining Method. IV. Transferring Coarse-Grained Potentials between Temperatures; *J. Chem. Phys.* **2009**, *131*, 024103.

[55] Das, A.; Andersen, H. C. The Multiscale Coarse-Graining Method. V. Isothermal-Isobaric Ensemble; *J. Chem. Phys.* **2010**, *132*, 164106.

[56] Hills, R. D.; Lu, L.; Voth, G. A. Multiscale Coarse-Graining of the Protein Energy Landscape; *PLoS Comput. Biol.* **2010**, *6*, e10000827.

[57] Larini, L.; Lu, L.; Voth, G. A. The Multiscale Coarse-Graining Method. VI. Implementation of Three-Body Coarse-Grained Potentials; *J. Chem. Phys.* **2010**, *132*, 164107.

[58] Sorenson, J. M.; Head-Gordon, T. Matching Simulation with Experiment: A New Simplified Model for Simulating Protein Folding; *J. Comput. Chem.* **2000**, *7*, 469–481.

[59] Tozzini, V.; Rocchia, W.; McCammon, J. A. Mapping All-Atom Models onto One-Bead Coarse-Grained Models: General Properties and Applications to a Minimal Polypeptide Model; *J. Chem. Theory Comput.* **2006**, *2*, 667–673.

[60] Liwo, A., et al. Cumulant-Based Expressions for the Multibody Terms for the Correlation between Local and Electrostatic Interactions in the United-Residue Force Field; *J. Chem. Phys.* **2001**, *115*, 2323–2347.

[61] Shen, H.; Liwo, A.; Scheraga, H. A. An Improved Functional Form for the Temperature Scaling Factors of the Components of the Mesoscopic UNRES Force Field for Simulations of Protein Structure and Dynamics; *J. Phys. Chem. B* **2009**, *113*, 8738–8744.

[62] Liwo, A., et al. Modification and Optimization of the United-Residue (UNRES) Potential Energy Function for Canonical Simulations. I. Temperature Dependence of the Effective Energy Function and Tests of the Optimization Method with Single Training Proteins; *J. Phys. Chem. B* **2007**, *111*, 260–285.

[63] Kubo, R. Generalized Cumulant Expansion Method; *J. Phys. Soc. Japan* **1962**, *17*, 1100–1120.

[64] Berne, B. J.; Pechukas. Gaussian Model Potential for Molecular Interactions; *J. Chem. Phys.* **1972**, *56* (8), 4213–4216.

[65] Kabadi, V. N. Molecular-Dynamics of Fluids: The Gaussian Overlap Model II; *Ber. Bunsenges. Phys. Chem.* **1986**, *90* (4), 327–332.

[66] Cleaver, D. J., et al. Extension and Generalization of the Gay–Berne Potential; *Phys. Rev. E. Stat. Phys. Plasmas Fluids Relat. Interdiscip. Topics* **1996**, *54* (1), 559–567.

[67] Leach, A. R. *Molecular Modelling: Principles and Applications,* 2nd ed.; Prentice Hall: Harlow, England; New York; xxiv, 744 p. [16] of plates.

[68] Applequist, J. Traceless Cartesian Tensor Forms for Spherical Harmonic-Functions - New Theorems and Applications to Electrostatics of Dielectric Media; *J. Phys. A: Math. Gen.* **1989**, *22* (20), 4303–4330.

[69] Bratko, D.; Blum, L.; Luzar, A. A Simple-Model for the Intermolecular Potential of Water; *J. Chem. Phys.* **1985**, *83* (12), 6367–6370.

[70] Liu, Y.; Ichiye, T. Soft Sticky Dipole Potential for Liquid Water: A New Model; *J. Phys. Chem.* **1996**, *100* (7), 2723–2730.

[71] Ichiye, T.; Tan, M. L. Soft Sticky Dipole-Quadrupole-Octupole Potential Energy Function for Liquid Water: An Approximate Moment Expansion; *J. Chem. Phys.* **2006**, *124* (13).

[72] Yesylevskyy, S. O., et al. Polarizable Water Model for the Coarse-Grained MARTINI Force Field; *PLoS Comput. Biol.* **2010**, *6*, e1000810.

[73] Wu, Z.; Cui, Q. A.; Yethiraj, A. A New Coarse-Grained Model for Water: The Importance of Electrostatic Interactions; *J. Phys. Chem. B* **2010**, *114* (32), 10524–10529.

[74] Hadley, K. R.; McCabe, C. On the Investigation of Coarse-Grained Models for Water: Balancing Computational Efficiency and the Retention of Structural Properties; *J. Phys. Chem. B* **2010**, *114* (13), 4590–4599.

[75] Basdevant, N.; Borgis, D.; Ha-Duong, T. A semi-implicit solvent model for the simulation of peptides and proteins; *J. Comput. Chem.* **2004**, *25* (8), 1015–1029.

[76] Basdevant, N.; Ha-Duong, T.; Borgis, D. Particle-Based Implicit Solvent Model for Biosimulations: Application to Proteins and Nucleic Acids Hydration; *J. Chem. Theory Comput.* **2006**, *2* (6), 1646–1656.

[77] Masella, M.; Borgis, D.; Cuniasse. Combining a Polarizable Force-Field and a Coarse-Grained Polarizable Solvent Model: Application to Long Dynamics Simulations of Bovine Pancreatic Trypsin Inhibitor; *J. Comput. Chem.* **2008**, *29* (11), 1707–1724.

[78] Ha-Duong, T.; Basdevant, N.; Borgis, D. A polarizable coarse-grained water model for coarse-grained proteins simulations; *Chem. Phys. Lett.* **2009**, *468* (1–3), 79–82.

[79] Masella, M.; Borgis, D.; Cuniasse. Combining a Polarizable Force-Field and a Coarse-Grained Polarizable Solvent Model. II. Accounting for Hydrophobic Effects; *J. Comput. Chem.* **2011**, *32* (12), 2664–2678.

[80] Feig, M.; Im, W.; Brooks, C. L. Implicit Solvation Based on Generalized Born theory in Different Dielectric Environments; *J. Chem. Phys.* **2004**, *120* (2), 903–911.

[81] Feig, M., et al. Performance Comparison of Generalized Born and Poisson Methods in the Calculation of Electrostatic Solvation Energies for Protein Structures; *J. Comput. Chem.* **2004**, *25* (2), 265–284.

[82] Hawkins, G. D.; Cramer, C. J.; Truhlar, D. G. Pairwise Solute Descreening of Solute Charges from a Dielectric Medium; *Chem. Phys. Lett.* **1995**, *246* (1–2), 122–129.

[83] Hawkins, G. D.; Cramer, C. J.; Truhlar, D. G. Parametrized Models of Aqueous Free Energies of Solvation Based on Pairwise Descreening of Solute Atomic Charges from a Dielectric Medium; *J. Phys. Chem.* **1996**, *100* (51), 19824–19839.

[84] Onufriev, A.; Bashford, D.; Case, D. A. Modification of the Generalized Born Model Suitable for Macromolecules; *J. Phys. Chem. B* **2000**, *104* (15), 3712–3720.

[85] Onufriev, A.; Bashford, D.; Case, D. A. Exploring Protein Native States and Large-Scale Conformational Changes with a Modified Generalized Born Model; *Proteins: Struct. Funct. Bioinf.* **2004**, *55* (2), 383–394.

[86] Onufriev, A.; Case, D. A.; Bashford, D. Effective Born Radii in the Generalized Born Approximation: The Importance of Being Perfect; *J. Comput. Chem.* **2002**, *23* (14), 1297–1304.

[87] Qiu, D., et al. The GB/SA Continuum Model for Solvation. A Fast Analytical Method for the Calculation Of Approximate Born Radii; *J. Phys. Chem. A* **1997**, *101* (16), 3005–3014.

[88] Schaefer, M.; Froemmel, C. A Precise Analytical Method for Calculating the Electrostatic Energy of Macromolecules in Aqueous-Solution; *J. Mol. Biol.* **1990**, *216* (4), 1045–1066.

[89] Schaefer, M.; Karplus, M. A Comprehensive Analytical Treatment of Continuum Electrostatics; *J. Phys. Chem.* **1996**, *100* (5), 1578–1599.

[90] Sigalov, G.; Fenley, A.; Onufriev, A. Analytical Electrostatics for Biomolecules: Beyond the Generalized Born Approximation; *J. Chem. Phys.* **2006**, *124* (12).

[91] Sigalov, G.; Scheffel; Onufriev, A. Incorporating Variable Dielectric Environments into the Generalized Born Model; *J. Chem. Phys.* **2005**, *122* (9).

[92] Still, W. C., et al. Semianalytical Treatment of Solvation for Molecular Mechanics and Dynamics; *J. Am. Chem. Soc.* **1990**, *112* (16), 6127–6129.

[93] Schnieders, M. J.; Ponder, J. W. Polarizable Atomic Multipole Solutes in a Generalized Kirkwood Continuum; *J. Chem. Theory Comput.* **2007**, *3* (6), 2083–2097.

[94] Ren, P. Y.; Ponder, J. W. Consistent Treatment of Inter- and Intramolecular Polarization in Molecular Mechanics Calculations; *J. Comput. Chem.* **2002**, *23* (16), 1497–1506.

[95] Ren, P. Y.; Ponder, J. W. Polarizable Atomic Multipole Water Model for Molecular Mechanics Simulation; *J. Phys. Chem. B* **2003**, *107* (24), 5933–5947.

[96] Makowski, M., et al. Simple Physics-Based Analytical Formulas for the Potentials of Mean Force for the Interaction of Amino Acid Side Chains in Water. 3. Calculation and Parameterization of the Potentials of Mean Force of Pairs of Identical Hydrophobic Side Chains; *J. Phys. Chem. B* **2007**, *111*, 2925–2931.

[97] Best, R. B.; Buchete, N. V.; Hummer, G. Are Current Molecular Dynamics Force Fields Too Helical? *Biophys. J.* **2008**, *95*, L7–L9.

[98] Hegefeld, W. A., et al. Helix Formation in a Pentapeptide Experiment and Forcefield Dependent Dynamics; *J. Phys. Chem. A* **2010**, *114*, 12391–12402.

[99] Anfinsen, C. B. Principles that Govern the Folding of Protein Chains; *Science* **1973**, *181*, 223–230.

[100] Ołdziej, S., et al. Optimization of the UNRES Force Field by Hierarchical Design of the Potential-Energy Landscape: III. Use of Many Proteins in Optimization; *J. Phys. Chem. B* **2004**, *108*, 16950–16959.

[101] Lee, J.; Scheraga, H. A.; Rackovsky, S. New Optimization Method for Conformational Energy Calculations on Polypeptides: Conformational Space Annealing; *J. Comput. Chem.* **1997**, *18*, 1222–1232.

[102] Chang, C. E., et al. Gated Binding of Ligands to HIV-1 Protease: Brownian Dynamics Simulations in a Coarse-Grained Model; *Biophys. J.* **2006**, *90*, 3880–3885.

[103] Chang, C. E., et al. Binding Pathways of Ligands to HIV-1 Protease: Coarse-Grained and Atomistic Simulations; *Chem. Biol. Drug Des.* **2007**, *65*, 5–13.

[104] Trylska, J., et al. HIV-1 Protease Substrate Binding and Product Release Pathway Explored with a Coarse-Grained Molecular Dynamics; *Biophys. J.* **2007**, *92*, 4179–4187.

Bioinformatics

Section Editor: Wei Wang

Department of Chemistry and Biochemistry, University of California-San Diego, La Jolla, CA, USA

CHAPTER SIX

Poisson–Boltzmann Implicit Solvation Models

Qin Cai*,†,1, Jun Wang†, Meng-Juei Hsieh†, Xiang Ye† and Ray Luo†
*Department of Biomedical Engineering, University of California, Irvine, CA, USA
†Department of Molecular Biology and Biochemistry, University of California, Irvine, CA, USA
1Corresponding author: E-mail: qcai@uci.edu

Contents

Abstract

Electrostatics plays a crucial role in determining the structures and functions of biomolecules. Implicit solvent models based on the Poisson–Boltzmann (PB) theory have enjoyed numerous successes in a variety of biological applications associated with electrostatics, especially in efficient calculations of solvation free energy and binding free energy. Efforts in adopting numerical PB methods for the more demanding dynamic simulations are also actively underway as more accurate numerical solvers and more powerful computers become available. Successful applications of these models certainly require a carefully parameterized force field that may still fail in challenging applications. There are also intrinsic limitations with implicit solvent models, which can also be resolved by explicit or semi-explicit representation of solvent molecules.

1. INTRODUCTION

Long-range electrostatic interactions play critical roles in determining the structures and functions of biomolecules [1,2]. In typical biochemical processes where a quantum mechanical treatment is not required, classical electrostatic modeling has been proved to be a useful modeling too, especially in applications involving highly charged biomolecules [3].

Annual Reports in Computational Chemistry, Volume 8
ISSN 1574-1400,
http://dx.doi.org/10.1016/B978-0-444-59440-2.00006-5
149

Biomolecules function in the aqueous environment. In biomolecular modeling, water molecules can be included either explicitly or implicitly, resulting in two categories of electrostatic models: the explicit solvent models and the implicit solvent models. In the explicit solvent models, the Coulomb's law is employed to compute pairwise electrostatic interactions over all explicitly represented atoms, with the number of water atoms apparently dominating most simulation systems. Due to the long-range nature of electrostatic interactions, a periodic boundary condition is often used to mimic a macroscopically large water box surrounding a solute molecule. With the periodic setup, the Ewald summation technique can be used to compute electrostatic energies and forces needed for molecular simulations [4].

Apparently in most biomolecular modeling studies, the biomolecular solute is of central interest. Thus, it is natural to explore alternative treatments of water where the water molecules are modeled implicitly. This allows the electrostatic interactions of the solvation system to be modeled with the Poisson–Boltzmann (PB) equation if the water is modeled as a high dielectric and the solute is modeled as a low dielectric [5]. In this regard, the solvation effect becomes a mean force, which is an average over all the solvent degrees of freedom [6]. If electrolytes are present, the PB model assumes that they obey a Boltzmann distribution at equilibrium. Finally, the electrostatic potential (ESP) across the space is characterized by the following PB equation

$$\nabla \cdot \varepsilon \nabla \phi = -4\pi \rho_0 - 4\pi \sum_i ez_i c_i \exp(-ez_i \phi / k_B T), \qquad (6.1)$$

where ε is the dielectric constant, ρ_0 is the solute charge density, e is the unit charge, z_i is the valence of ion type i, c_i is the number density of ion type i, k_B is the Boltzmann constant and T is the absolute temperature. If the ESP is weak and the ionic strength is low [7], the above nonlinear PB equation can be simplified to a linear PB equation

$$\nabla \cdot \varepsilon \nabla \phi = -4\pi \rho_0 + \varepsilon_v \kappa^2 \phi, \qquad (6.2)$$

where $\kappa^2 = 8\pi e^2 I / \varepsilon_v k_B T$ and $C = ez / k_B T$. Here, v denotes the solvent and I represents the ionic strength of the solution and is computed as $I = z^2 c$.

Due to the considerable computational expense in solving the PB equation, simpler methods have also been developed for the continuum electrostatics modeling of solvation, such as the induced multipole model [8], the analytical continuum methods [9], the dielectric screening model [10], and the generalized Born (GB) model [11,12]. The GB model has received universal attention and demonstrated its advantages in dynamic simulations [13].

Many variations exist in the underlying model setup based on different approximations, leading to different solutions of the PB Eqn (6.1). Choices can be made in solute charge treatment, space boundary condition, solute–solvent interface location, dielectric determination, and so on. Some of these issues will be discussed in the Section 6.3.

Point-charge models are prevalent in representing atomic charges ρ_0 in Eqn (6.1). The resultant singularity of the delta function of ρ_0 may cause difficulty to numerical solvers of the PB equation [14]. The possible remedy is the regularization of Eqn (6.1), and a common strategy in regularization is to decompose the solution into regular component and singular component and focus on the former in the numerical solutions [15]. Apparently, it takes extra time to compute the singular component analytically, but it benefits the convergence and save additional computing time in a reference system for reaction energy calculations [16].

It is apparently impossible and unnecessary to solve Eqn (6.1) in an infinite space. Thus, a space boundary has to be set up properly. To mimic the free space solution, the boundary needs to be set adequately far away from the solute for accurate approximation. A widely used space boundary condition is the Debye–Hückel approximation

$$\phi^{BC} = \frac{1}{\varepsilon_v} \sum_i \frac{Q_i \exp[-\kappa(R_i - r_i)]}{R_i(1 + \kappa r_i)}, \tag{6.3}$$

where $\sum_i$ is a sum over all atoms, Q_i is the charge of the nth atom, r_i is the radius of the ith atom, and R_i is the distance between the ith atom and the center of the grid. Equation (6.3) is derived from the solution to the linear PB equation of a single sphere immersed in an electrolytic solution, but it is not applicable to the nonlinear PB equation in certain circumstances [17,18]. The PB model can also be used to study membrane proteins, where a periodic boundary condition in the membrane plane has to be employed [19].

2. BIOLOGICAL APPLICATIONS

The PB model has a variety of biological applications. The pK_a value of an ionizable group is determined by electrostatic interactions and thus can be calculated with the PB model [20–23]. The ionizable group alone in solution is used as a reference state, and the change in the electrostatic energy due to the change in the environment is evaluated to predict the pK_a value of the same group in a protein [24]. Such pK_a calculations play an important role in the setup of biomolecular simulations, the investigation of ligand binding and catalysis, and the identification of enzyme active sites [25]. The constant pH dynamics with the PB implicit solvent was also proposed to improve the pK_a calculations [26–29].

The PB model has also been used for estimation of solvation free energies [13,30] and binding free energies [31–33]. In particular, the Molecular Mechanics/Poisson Boltzmann Surface Area (MM/PBSA) method is a very useful tool to calculate the binding free energies at low computational cost compared to thermodynamic integration and free energy perturbation with the explicit solvent model [34]. In this method, a molecular dynamic simulation of the biomolecular system of interest is carried out and then the snapshots of the simulation are processed with the PB method to estimate the conformational free energies [35]. For the study of binding affinity, the free energy difference between separation and association of two molecules is calculated often from a single trajectory of the complex because no large structural changes are expected upon binding [34]. Brown et al. implemented a high-throughput version of MM/PBSA with comparable accuracy with the original approach but with dramatically reduced computational time. Two orders of magnitude reduction were observed for virtual screening applications [36].

A direct benefit of the PB model is that the ESP profile becomes available for visualization right after the PB equation is solved. Coloring the molecular surface by visualization tools, such as VMD [37], GRASP [38], and PyMOL [39], according to the solution to the PB equation provides us with insight of the electrostatic properties of biomolecules for structural analysis [40]. The PBEQ solver also provides a web-based graphical user interface to read biomolecular structures, solve the PB equations and interactively visualize the ESP [41].

The PB model has the great potential of aiding peptide or protein design. Marshall et al. presented a revised finite-difference PB method with reduced representation of protein surface so that the electrostatic energy becomes pairwise decomposable by side chains and compatible with protein design calculations [42]. Kieslich et al. recently developed a computational framework known as Analysis of Electrostatic Similarity Of Proteins (AESOP) based on the PB electrostatics [43] and applied this framework to the design of mutant proteins with enhanced immunological activity [44].

3. FORCE FIELD CONSIDERATIONS

The quality in the analyses or predictions with implicit solvent models is highly dependent on the model parameters or the force fields, which include the atomic radii, the atomic partial charges, and the dielectric constants. Ideally, we should directly compare implicit solvent simulations with experiment to assess the quality of implicit solvent models. However, such comparisons are often limited by several factors: the inability to generate convergent ensembles of conformations so that the error in free energy due to lack of entropy terms can be as large as 2 kcal/mol [45]; the coupling between electrostatic solvation and nonelectrostatic solvation treatments so that it is hard to interpret the disagreement with experiment if any; the limited accuracy in the classic molecular mechanics force fields again makes it hard to interpret the disagreement with experiment if any; and finally, the molecules are limited to the small and neutral ones for which experimental results are available. Thus, a more straightforward question to ask is how well implicit solvents agree with explicit solvents under identical simulation conditions.

It is well known that PB reaction field energies depend sensitively on the solute–solvent interface and the dielectric function [46–49] so that different reaction field energies computed with different cavity radii may have different performances. To date, several sets of systemically optimized cavity radii have been presented [49–52]. An important issue that has to be addressed is the transferability of optimized cavity radii from small molecules (usually in the training set) to large molecules (usually biomolecules out of the training set). However, transferability tests, i.e. tests of cavity radii with biomolecules outside the training set, were only mentioned in Swanson et al. [49] It is found that transferability of cavity radii cannot be taken for granted, as shown for two sets of radii with the n-methylamine (NMA) dimer as a test case [53].

After the atomic radii have been chosen, the interface can be determined with well-known hard sphere surfaces, such as the van der Waals surface (VDWS), the solvent-accessible surface (SAS) [54], and the solvent-excluded surface (SES) [55]. The VDWS definition tends to have solvent pocket within the solute molecule, while the SAS, or the extended VDWS, is difficult to reproduce the electrostatic energetics in the explicit solvent models due to its much enlarged atomic cavities. Although the PB methods with the SES is able to reproduce solvation energy and forces from explicit solvent models [56], the SES is not differentiable and may cause instability in dynamics [57]. All of the above hard sphere surfaces have cusps as well as a sharp transition from low dielectric to high dielectric. Harmonic averaging of the solute and solvent dielectric constants is a simple and effective way to smooth the transition and improve the convergence of energy and force calculations [58]. A smoother Gaussian surface [47] and a spline-based dielectric model [56] were also proposed. In this type of approach, a distance-dependent density/volume exclusion function is used to define each atomic volume. However, the cost of computing the volume exclusion/density function at every grid edge is a major concern when this approach is applied to numerical solvers. In addition, these atom-centered surfaces produce large volumes of interstitial high dielectrics in globular proteins, which artificially overestimate solvation energies [59]. Considering these limitations, the modified VDWS was proposed to improve the PB methods [57].

Solute permanent dipoles are represented by atomic partial charges in commonly used force fields, and they are often determined from *ab initio* gas-phase calculations on small compounds. The charge sets are obtained from the computed electronic densities by fitting to ESPs [60,61]. To overcome the drawback of the ESP methods that different conformations may produce considerable variations, a restrained ESP-fit (RESP) model was developed [62]. A final scaling procedure may be applied to correct for polarization effects arising in the condensed phase [63,64]. As the computational power grows continuously, the polarization effects in the condensed phase can be more accurately captured by explicitly incorporating induced dipoles or even higher order multipoles, in polarizable force fields, such as those in OPLS-AA [65], CHARMM [66,67], AMOEBA [68], and AMBER [69,70].

Before polarizable force fields mature and their behavior in free energy calculations become completely understood, the point-charge models still dominate the field. In these simplified models, the solute dielectric constant can be used to account for the solute polarization effect. In reality, the solute dielectric constant is not universal or transferrable and has been reported to be ranging from 2 to 40, depending on the system and application [71,72]. Gilson et al. used the Kirkwood–Fröhlich dielectric theory to compute the dielectric constant for a folded protein, which falls between 2.5 and 4 [73]. In contrast, another study with the same theory on the dipole fluctuations observed in MD simulations resulted in a large average dielectric constants of 15–40 due to motions of charged side chains, suggesting the spatial heterogeneity of protein dielectric responses: low dielectric value in the interior and high dielectric value on the surface [74]. That is probably why pK_a calculations often require a relatively high solute dielectric constants [75,76], whereas in other types of calculations like binding affinity, a smaller dielectric constant is more appropriate [77,78]. Attempts have also been made to model the heterogeneity of the solute dielectric by using distance-dependent dielectric model [79] or nonuniform charge scaling [80], which are efficient but empirical and difficult to interpret.

4. NUMERICAL SOLVERS

Currently, there are three major numerical methods widely used for solving the PB equation. One such approach is the finite-difference method (FDM) [81–85]. In this method, the physical properties of the solution, such as the charge density and dielectric constant, are mapped onto a cubic or rectangular lattice, and a discrete approximation to the governing partial differential equations is produced. With the regularization scheme adopted, the significant error in the potential nearby the point charges is eliminated. The major issue is the dramatic change in the potential close to the interface, which can be alleviated with an adaptive Cartesian grid [86,87]. Alternatively, additional equations are employed to enforce the boundary conditions exactly on the interface, as used in the matched interface and boundary (MIB) [88] method and immersed inter-face method (IIM) [89].

The second approach is the finite-element method (FEM) [14,90–93], which is based on the weak variational formulation. The potential to be solved is approximated by a superposition of a set of basis functions. A linear or nonlinear system for the coefficients produced by the weak formulation has to be solved. A nice property of the FEM is that the mesh is unstructured so that adaptivity can be achieved. Also body-fitted mesh can be generated to fit nicely to the boundary or the interface, such as the molecular surface. However, this strategy may cause additional cost, i.e. constant remeshing, for dynamical problems.

The third approach is the boundary-element method (BEM) [94–100]. In BEM, the Poisson or PB equation is solved for either the induced surface charge [94,97,100] or the normal component of the electric displacement [95,96,98,99] on the dielectric boundary between the solute and the solvent. A surface mesh generation program is necessary to discretize the solute–solvent interface to implement the BEM solver. MSMS is widely used because of its high efficiency [101]. Chen et al. recently developed a robust program named TMSmesh, capable of handling biomolecules consisting of more than 1 million atoms [102]. A combination of FDM and BEM [103] was also reported.

Some of these methods have been incorporated into the popular PB programs, including Delphi [84], UHBD [82], PBEQ [56], PBSA [85], and APBS [90]. Fenley et al. proposed a correlated Monte Carlo approach to solve the PB equation. The most appealing feature of this method is that the CPU time is on a logarithmic scaling of the number of atoms [104].

5. DISTRIBUTIVE AND PARALLEL IMPLEMENTATION

The iterative methods of numerical PB solvers can be loosely grouped into two types: stationary methods and Krylov subspace methods. The partition is not very strict because many stationary methods can be used in preconditioning for the Krylov subspace methods [105]. To implement stationary methods, such as Jacobi, successive over-relaxation, and Gauss–Seidel, for distributive computing environments, the unknowns are usually reordered with multicoloring [106] or multisplitting approaches [107].

To implement Krylov subspace methods on distributive computing environments, possible strategies are domain decompositions [108–110] and distributive preconditioners [111,112]. Hsieh et al. have explored the distributive multiblock focusing technique to analyze what it takes to achieve an acceptable accuracy and good performance in numerical PB solutions of large biomolecular systems. A highly scalable parallel version of the method is also implemented and incorporated into the Amber/PBSA program [113].

Other efforts to solve linear problems include the use of two or more-level grid-based solution such as the multigrid methods [90,114]. A seemingly related method is electrostatic focusing in PB. The focusing technique is a series of finite-difference runs that are performed with successively finer grids, each run having boundary conditions calculated from the potential map of its predecessor [115]. A limitation of the electrostatic focusing calculations is that the memory usage of the finest grid for the region of interest may still be unmanageably large in PB calculations of large biomolecular systems. It has been reported that the fine grids in the focusing technique can be divided into multiple smaller blocks for distributive computation [116].

6. FORCE AND DYNAMICS CALCULATIONS

With optimized algorithms and parallel implementation of PB solvers, it is possible to extend the PB applications to molecular dynamic simulations. Indeed, pioneer work has been done in this field [57,117–119]. All of them addressed the continuity of the dielectric constant on the solute–solvent interface and used smoothing techniques. The classical SES definition is not differentiable over time [47] and therefore either a Gaussian surface or a modified VDWS was used. However, the SES probably gives the best agreement with explicit solvents and experiment [59,120,121]. Thus, revision of dielectric models for implicit solvents has to balance both dynamic stability and quality [122].

Another concern related to the PB molecular dynamics is the problem of assigning solvation forces. It has been shown that solvation forces can be divided into three components: reaction field forces, dielectric boundary forces (DBF) and ionic boundary forces [123]. Reaction field forces are straightforward to compute, ionic boundary forces are generally small and can be neglected, but accurate computation of DBF is a challenge [46]. The DBF formulations proposed by Gilson et al. and Im et al. require smoothly varying dielectric constants on the interface [56,123], while the DBF formulation proposed by Cai et al. is suitable for classical two-dielectric model with harmonic averaging [124]. Recently, Li et al. revisited the DBF calculation through a variational strategy in the classical two-dielectric model [125]. An equivalent DBF formulation is also available for the BEM [126] and FEM [127] solvers.

7. LIMITATIONS

In the studies of comparison between the PB model and the explicit water model, water-mediated salt bridging is one of the major discrepancies, although overall agreement between the two models can be observed [53,128]. This problem can hardly be solved unless explicit water molecules are present surrounding the solute in the model, and therefore, the implicit/explicit model has received attention [129]. More research work has been conducted on the effects of ion size and a modified PB model has been accordingly used to include these effects [130,131].

Mobley et al. studied the hydration behavior of polar solutes and found that due to the asymmetry of water charge centers, solutes having a large negative charge balancing diffuse positive charges are preferentially solvated relative to those having a large positive charge balancing diffuse negative charges. They pointed out that such effects cannot be captured by implicit solvent models, which respond symmetrically with respect to charge [132].

The PB model has long been criticized for its unreliable estimate of the entropy due to the solute conformational change, which is often done with a quasi-harmonic analysis or normal mode analysis [133]. However, the entropy estimates by the quasi-harmonic analysis is hard to converge and probably applicable only to small conformational changes [133,134]. In addition, the deviations of the estimates are large, especially for those receptors binding polar drug-like molecules [135]. The deviation can be reduced after a buffer region is included in the normal mode analysis [136]. Wittayanarakul et al. found that inclusion of explicit water molecules in the PB model can significantly improve the entropy estimate and the overall absolute binding energy [137]. Under such circumstance, the MM/PBSA method has better performance in analyzing relative binding energy of similar ligands [34].

Despite the aforementioned PB molecular dynamics studies, those successful attempts could only apply such solvents in simulations to small organic molecules since the cost of using implicit solvents on macromolecules was prohibitively high. In fact, the per-step simulation cost is higher with the finite-difference approach than that with explicit water even at the 1 Å grid spacing, though it can be argued that the solvent is always equilibrated in implicit solvents, whereas it takes a long time to achieve equilibration in explicit water simulations. Its inefficiency sharply limits the practical applications of implicit solvents in routine dynamic simulations of macromolecules.

REFERENCES

[1] Davis, M. E.; McCammon, J. A. *Chem. Rev.* **1990**, *90*, 509.
[2] Perutz, M. F. *Science* **1978**, *201*, 1187.
[3] Honig, B.; Nicholls, A. *Science* **1995**, *268*, 1144.
[4] Essmann, U.; Perera, L.; Berkowitz, M. L.; Darden, T.; Lee, H.; Pedersen, L. G. *J. Chem. Phys.* **1995**, *103*, 8577.
[5] Warwicker, J.; Watson, H. C. *J. Mol. Biol.* **1982**, *157*, 671.

[6] Wang, J.; Tan, C. H.; Tan, Y. H.; Lu, Q.; Luo, R. *Commun. Comput. Phys.* **2008**, *3*, 1010.
[7] Hill, T. L. Dilute Electrolyte Solutions and Plasmas In An Introduction to Statistical Thermodynamics; Dover Publications, Inc.: New York; **1986** p 321.
[8] Davis, M. E. *J. Chem. Phys.* **1994**, *100*, 5149.
[9] Schaefer, M.; Karplus, M. *J. Phys. Chem.* **1996**, *100*, 1578.
[10] Luo, R.; Moult, J.; Gilson, M. K. *J. Phys. Chem. B* **1997**, *101*, 11226.
[11] Still, W. C.; Tempczyk, A.; Hawley, R. C.; Hendrickson, T. *J. Am. Chem. Soc.* **1990**, *112*, 6127.
[12] Onufriev, A.; Case, D. A.; Bashford, D. *J. Comput. Chem.* **2002**, *23*, 1297.
[13] Shivakumar, D.; Deng, Y.; Roux, B. *J. Chem. Theory Comput.* **2009**, *5*, 919.
[14] Chen, L.; Holst, M. J.; Xu, J. C. *SIAM J. Numer. Anal.* **2007**, *45*, 2298.
[15] Lu, B. Z.; Zhou, Y. C.; Holst, M. J.; McCammon, J. A. *Commun. Comput. Phys.* **2008**, *3*, 973.
[16] Cai, Q.; Wang, J.; Zhao, H. K.; Luo, R. *J. Chem. Phys.* **2009**, *130*, 145101.
[17] Rocchia, W. *Math. Comput. Model.* **2005**, *41*, 1109.
[18] Boschitsch, A. H.; Fenley, M. O. *J. Comput. Chem.* **2007**, *28*, 909.
[19] Sayyed-Ahmad, A.; Kaznessis, Y. N. *Plos One* **2009**, *4*.
[20] Georgescu, R. E.; Alexov, E. G.; Gunner, M. R. *Biophys. J.* **2002**, *83*, 1731.
[21] Nielsen, J. E.; McCammon, J. A. *Protein Sci.* **2003**, *12*, 313.
[22] Warwicker, J. *Protein Sci.* **2004**, *13*, 2793.
[23] Tang, C. L.; Alexov, E.; Pyle, A. M.; Honig, B. *J. Mol. Bio.* **2007**, *366*, 1475.
[24] Alexov, E.; Mehler, E. L.; Baker, N.; Baptista, A. M.; Huang, Y.; Milletti, F.; Nielsen, J. E.; Farrell, D.; Carstensen, T.; Olsson, M. H. M.; Shen, J. K.; Warwicker, J.; Williams, S.; Word, J. M. *Proteins* **2011**, *79*, 3260.
[25] Baker, N. A. *Method Enzymol.* **2004**, *383*, 94.
[26] Dlugosz, M.; Antosiewicz, J. M.; Robertson, A. D. *Phys. Rev. E* **2004**, *69*, 021915.
[27] Dlugosz, M.; Antosiewicz, J. M. *J. Phys. Chem. B* **2005**, *109*, 13777.
[28] Dlugosz, M.; Antosiewicz, J. M. *Chem. Phys.* **2004**, *302*, 161.
[29] Machuqueiro, M.; Baptista, A. M. *J. Phys. Chem. B* **2006**, *110*, 2927.
[30] Nicholls, A.; Mobley, D. L.; Guthrie, J. P.; Chodera, J. D.; Bayly, C. I.; Cooper, M. D.; Pande, V. S. *J. Med. Chem.* **2008**, *51*, 769.
[31] Swanson, J. M. J.; Henchman, R. H.; McCammon, J. A. *Biophys. J.* **2004**, *86*, 67.
[32] Bertonati, C.; Honig, B.; Alexov, E. *Biophys. J.* **2007**, *92*, 1891.
[33] Brice, A. R.; Dominy, B. N. *J. Comput. Chem.* **2011**, *32*, 1431.
[34] Homeyer, N.; Gohlke, H. *Mol. Inform.* **2012**, *31*, 114.
[35] Kollman, P. A.; Massova, I.; Reyes, C.; Kuhn, B.; Huo, S. H.; Chong, L.; Lee, M.; Lee, T.; Duan, Y.; Wang, W.; Donini, O.; Cieplak, P.; Srinivasan, J.; Case, D. A.; Cheatham, T. E. *Acc. Chem. Res.* **2000**, *33*, 889.
[36] Brown, S. P.; Muchmore, S. W. *J. Chem. Inform. Model.* **2007**, *47*, 1493.
[37] Humphrey, W.; Dalke, A.; Schulten, K. *J. Mol. Graph. Model.* **1996**, *14*, 33.
[38] Nicholls, A.; Sharp, K. A.; Honig, B. *Proteins* **1991**, *11*, 281.
[39] The PyMOL Molecular Graphics System, Version 1.5.0.4, Schrödinger, LLC.
[40] Yang, X. L.; Xie, J.; Niu, B.; Hu, X. N.; Gao, Y.; Xiang, Q.; Zhang, Y. H.; Guo, Y.; Zhang, Z. G. *J. Mol. Graph. Model.* **2005**, *23*, 389.
[41] Jo, S.; Vargyas, M.; Vasko-Szedlar, J.; Roux, B.; Im, W. *Nucleic Acids Res.* **2008**, *36*, W270.
[42] Marshall, S. A.; Vizcarra, C. L.; Mayo, S. L. *Protein Sci.* **2005**, *14*, 1293.
[43] Kieslich, C. A.; Morikis, D.; Yang, J.; Gunopulos, D. *Biotechnol. Prog.* **2011**, *27*, 316.
[44] Pyaram, K.; Kieslich, C. A.; Yadav, V. N.; Morikis, D.; Sahu, A. *J. Immunol.* **2010**, *184*, 1956.
[45] Mobley, D. L.; Dill, K. A.; Chodera, J. D. *J. Phys. Chem. B* **2008**, *112*, 938.
[46] Wagoner, J.; Baker, N. A. *J. Comput. Chem.* **2004**, *25*, 1623.
[47] Grant, J. A.; Pickup, B. T.; Nicholls, A. *J. Comput. Chem.* **2001**, *22*, 608.

[48] Nina, M.; Im, W.; Roux, B. *Biophys. Chem.* **1999**, *78*, 89.

[49] Swanson, J. M. J.; Adcock, S. A.; McCammon, J. A. *J. Chem. Theory Comput.* **2005**, *1*, 484.

[50] Banavali, N. K.; Roux, B. *J. Phys. Chem. B* **2002**, *106*, 11026.

[51] Nina, M.; Beglov, D.; Roux, B. *J. Phys. Chem. B* **1997**, *101*, 5239.

[52] Sitkoff, D.; Sharp, K. A.; Honig, B. *J. Phy. Chem.* **1994**, *98*, 1978.

[53] Tan, C. H.; Yang, L. J.; Luo, R. *J. Phys. Chem. B* **2006**, *110*, 18680.

[54] Lee, B.; Richards, F. M. *J. Mol. Biol.* **1971**, *55*, 379.

[55] Richards, F. M. *Annu. Rev. Biophys. Bioeng.* **1977**, *6*, 151.

[56] Im, W.; Beglov, D.; Roux, B. *Comput. Phys. Commun.* **1998**, *111*, 59.

[57] Lu, Q.; Luo, R. *J. Chem. Phys.* **2003**, *119*, 11035.

[58] Davis, M. E.; McCammon, J. A. *J. Comput. Chem.* **1991**, *12*, 909.

[59] Swanson, J. M. J.; Mongan, J.; McCammon, J. A. *J. Phys. Chem. B* **2005**, *109*, 14769.

[60] Weiner, S. J.; Kollman, P. A.; Case, D. A.; Singh, U. C.; Ghio, C.; Alagona, G.; Profeta, S.; Weiner, P. *J. Am. Chem. Soc.* **1984**, *106*, 765.

[61] Weiner, S. J.; Kollman, P. A.; Nguyen, D. T.; Case, D. A. *J. Comput. Chem.* **1986**, *7*, 230.

[62] Bayly, C. I.; Cieplak, P.; Cornell, W. D.; Kollman, P. A. *J. Phys. Chem.* **1993**, *97*, 10269.

[63] Cornell, W. D.; Cieplak, P.; Bayly, C. I.; Gould, I. R.; Merz, K. M.; Ferguson, D. M.; Spellmeyer, D. C.; Fox, T.; Caldwell, J. W.; Kollman, P. A. *J. Am. Chem. Soc.* **1995**, *117*, 5179.

[64] Duan, Y.; Wu, C.; Chowdhury, S.; Lee, M. C.; Xiong, G. M.; Zhang, W.; Yang, R.; Cieplak, P.; Luo, R.; Lee, T.; Caldwell, J.; Wang, J. M.; Kollman, P. *J. Comput. Chem.* **2003**, *24*, 1999.

[65] Jorgensen, W. L.; Maxwell, D. S.; TiradoRives, J. *J. Am. Chem. Soc.* **1996**, *118*, 11225.

[66] Patel, S.; Brooks, C. L. *J. Comput. Chem.* **2004**, *25*, 1.

[67] Patel, S.; Mackerell, A. D.; Brooks, C. L. *J. Comput. Chem.* **2004**, *25*, 1504.

[68] Schnieders, M. J.; Baker, N. A.; Ren, P. Y.; Ponder, J. W. *J. Chem. Phys.* **2007**, *126*, 124114.

[69] Wang, J.; Cieplak, P.; Li, J.; Hou, T.; Luo, R.; Duan, Y. *J. Phys. Chem. B* **2011**, *115*, 3091.

[70] Wang, J.; Cieplak, P.; Li, J.; Wang, J.; Cai, Q.; Hsieh, M.; Lei, H.; Luo, R.; Duan, Y. *J. Phys. Chem. B* **2011**, *115*, 3100.

[71] Kukic, P.; Nielsen, J. E. *Future Med. Chem.* **2010**, *2*, 647.

[72] Sheinerman, F. B.; Norel, R.; Honig, B. *Curr. Opin. Struct. Biol.* **2000**, *10*, 153.

[73] Gilson, M. K.; Honig, B. H. *Biopolymers* **1986**, *25*, 2097.

[74] Pitera, J. W.; Falta, M.; van Gunsteren, W. F. *Biophys. J.* **2001**, *80*, 2546.

[75] Antosiewicz, J.; McCammon, J. A.; Gilson, M. K. *J. Mol. Biol.* **1994**, *238*, 415.

[76] Nielsen, J. E.; Vriend, G. *Proteins* **2001**, *43*, 403.

[77] Yang, T.; Wu, J. C.; Yan, C.; Wang, Y.; Luo, R.; Gonzales, M. B.; Dalby, K. N.; Ren, P. *Proteins* **2011**, *79*, 1940.

[78] Hou, T. J.; Wang, J. M.; Li, Y. Y.; Wang, W. *J. Chem. Inform. Model.* **2011**, *51*, 69.

[79] Morozov, A. V.; Kortemme, T.; Baker, D. *J. Phys. Chem. B* **2003**, *107*, 2075.

[80] Schwarzl, S. M.; Huang, D. Z.; Smith, J. C.; Fischer, S. *J. Comput. Chem.* **2005**, *26*, 1359.

[81] Davis, M. E.; McCammon, J. A. *J. Comput. Chem.* **1989**, *10*, 386.

[82] Luty, B. A.; Davis, M. E.; McCammon, J. A. *J. Comput. Chem.* **1992**, *13*, 1114.

[83] Holst, M.; Saied, F. *J. Comput. Chem.* **1993**, *14*, 105.

[84] Rocchia, W.; Alexov, E.; Honig, B. *J. Phys. Chem. B* **2001**, *105*, 6507.

[85] Luo, R.; David, L.; Gilson, M. K. *J. Comput. Chem.* **2002**, *23*, 1244.

[86] Boschitsch, A. H.; Fenley, M. O. *J. Chem. Theory Comput.* **2011**, *7*, 1524.

[87] Mirzadeh, M.; Theillard, M.; Gibou, F. *J. Comput. Phys.* **2011**, *230*, 2125.

[88] Chen, D.; Chen, Z.; Chen, C.; Geng, W.; Wei, G.-W. *J. Comput. Chem.* **2011**, *32*, 756.

[89] Wang, J.; Cai, Q.; Li, Z. L.; Zhao, H. K.; Luo, R. *Chem. Phys. Lett.* **2009**, *468*, 112.

[90] Holst, M.; Baker, N.; Wang, F. *J. Comput. Chem.* **2000**, *21*, 1319.
[91] Shestakov, A. I.; Milovich, J. L.; Noy, A. *J. Colloid Interface Sci.* **2002**, *247*, 62.
[92] Bond, S. D.; Chaudhry, J. H.; Cyr, E. C.; Olson, L. N. *J. Comput. Chem.* **2010**, *31*, 1625.
[93] Xie, D.; Zhou, S. *Bit Numer. Math.* **2007**, *47*, 853.
[94] Zauhar, R. J.; Morgan, R. S. *J. Comput. Chem.* **1988**, *9*, 171.
[95] Yoon, B. J.; Lenhoff, A. M. *J. Comput. Chem.* **1990**, *11*, 1080.
[96] Juffer, A. H.; Botta, E. F. F.; Vankeulen, B. A. M.; Vanderploeg, A.; Berendsen, H. J. C. *J. Comput. Phys.* **1991**, *97*, 144.
[97] Vorobjev, Y. N.; Scheraga, H. A. *J. Comput. Chem.* **1997**, *18*, 569.
[98] Boschitsch, A. H.; Fenley, M. O.; Zhou, H. X. *J. Phys. Chem. B* **2002**, *106*, 2741.
[99] Lu, B. Z.; Cheng, X. L.; Huang, J. F.; McCammon, J. A. *Proc. Natl. Acad. Sci. USA* **2006**, *103*, 19314.
[100] Bardhan, J. P. *J. Chem. Phys.* **2009**, *130*, 094102.
[101] Sanner, M. F.; Olson, A. J.; Spehner, J. C. *Biopolymers* **1996**, *38*, 305.
[102] Chen, M.; Lu, B. *J. Chem. Theory Comput.* **2011**, 7, 203.
[103] Boschitsch, A. H.; Fenley, M. O. *J. Comput. Chem.* **2004**, *25*, 935.
[104] Fenley, M. O.; Mascagni, M.; McClain, J.; Silalahi, A. R. J.; Simonov, N. A. *J. Chem. Theory Comput.* **2010**, *6*, 300.
[105] Strang, G. Iterative Method for Ax = b In Linear Algebra and Its Applications. Thomson Brooks/Cole. **1988** p 380.
[106] Evans, D. J. *Parallel Comput.* **1984**, *1*, 3.
[107] Oleary, D. P.; White, R. E. *SIAM J. Alg. Disc. Meth.* **1985**, *6*, 630.
[108] Chen, W.; Poirier, B. *J. Comput. Phys.* **2006**, *219*, 185.
[109] Chen, W.; Poirier, B. *J. Comput. Phys.* **2006**, *219*, 198.
[110] Bank, R. E.; Holst, M. *SIAM J. Sci. Comput.* **2000**, *22*, 1411.
[111] Ma, S. *IEICE Trans. Fund. Electr. Commun. Comput. Sci.* **2008**, *E91A*, 2578.
[112] Iwashita, T.; Shimasaki, M. *IEEE Trans. Magn.* **2002**, *38*, 429.
[113] Hsieh, M.-J.; Luo, R. *J. Mol. Model.* **2011**, *17*, 1985.
[114] Baker, N.; Holst, M.; Wang, F. *J. Comput. Chem.* **2000**, *21*, 1343.
[115] Klapper, I.; Hagstrom, R.; Fine, R.; Sharp, K.; Honig, B. *Proteins* **1986**, *1*, 47.
[116] Baker, N. A.; Sept, D.; Joseph, S.; Holst, M. J.; McCammon, J. A. *Proc. Nat. Acad. Sci. USA* **2001**, *98*, 10037.
[117] Fogolari, F.; Brigo, A.; Molinari, H. *Biophys. J.* **2003**, *85*, 159.
[118] Prabhu, N. V.; Zhu, P. J.; Sharp, K. A. *J. Comput. Chem.* **2004**, *25*, 2049.
[119] Wang, J.; Tan, C. H.; Chanco, E.; Luo, R. *Phys. Chem. Chem. Phys.* **2010**, *12*, 1194.
[120] Lee, M. S.; Olson, M. A. *J. Phys. Chem. B* **2005**, *109*, 5223.
[121] Rocchia, W.; Sridharan, S.; Nicholls, A.; Alexov, E.; Chiabrera, A.; Honig, B. *J. Comput. Chem.* **2002**, *23*, 128.
[122] Chocholousova, J.; Feig, M. *J. Comput. Chem.* **2006**, *27*, 719.
[123] Gilson, M. K.; Davis, M. E.; Luty, B. A.; McCammon, J. A. *J. Phys. Chem.* **1993**, *97*, 3591.
[124] Cai, Q.; Ye, X.; Wang, J.; Luo, R. *Chem. Phys. Lett.* **2011**, *514*, 368.
[125] Li, B.; Cheng, X.; Zhang, Z. *SIAM J. Appl. Math.* **2011**, *71*, 2093.
[126] Zauhar, R. J. *J. Comput. Chem.* **1991**, *12*, 575.
[127] Cortis, C. M.; Friesner, R. A. *J. Comput. Chem.* **1997**, *18*, 1591.
[128] Salari, R.; Chong, L. T. *J. Phys. Chem. Lett.* **2010**, *1*, 2844.
[129] Lin, Y.; Baumketner, A.; Deng, S.; Xu, Z.; Jacobs, D.; Cai, W. *J. Chem. Phys.* **2009**, *131*.
[130] Silalahi, A. R. J.; Boschitsch, A. H.; Harris, R. C.; Fenley, M. O. *J. Chem. Theory Comput.* **2010**, *6*, 3631.
[131] Zhou, S.; Wang, Z.; Li, B. *Phys. Rev. E* **2011**, *84*.
[132] Mobley, D. L.; Barber, A. E., II; Fennell, C. J.; Dill, K. A. *J. Phys. Chem. B* **2008**, *112*, 2405.

[133] Srinivasan, J.; Cheatham, T. E.; Cieplak, P.; Kollman, P. A.; Case, D. A. *J. Am. Chem. Soc.* **1998**, *120*, 9401.

[134] Gohlke, H.; Case, D. A. *J. Comput. Chem.* **2004**, *25*, 238.

[135] Kongsted, J.; Soderhjelm, P.; Ryde, U. *J. Comput. Aided Mol. Des.* **2009**, *23*, 395.

[136] Kongsted, J.; Ryde, U. *J. Comput. Aided Mol. Des.* **2009**, *23*, 63.

[137] Wittayanarakul, K.; Hannongbua, S.; Feig, M. *J. Comput. Chem.* **2008**, *29*, 673.

SUBJECT INDEX

CUMULATIVE INDEX